AF615124

APPLIED STATISTICAL SCIENCE, II

APPLIED STATISTICAL SCIENCE, II

M. AHSANULLAH
EDITOR

PAPERS IN HONOR OF
MUNIR AHMAD

NOVA SCIENCE PUBLISHERS, INC.
COMMACK, NEW YORK

Creative Design: Gavin Aghamore
Editorial Production: Susan Boriotti
Assistant Vice President/Art Director: Maria Ester Hawrys
Office Manager: Annette Hellinger
Graphics: Frank Grucci
Manuscript Coordinator: Phyllis Gaynor
Book Production: Michelle Keller, Ludmila Kwartiroff, Christine Mathosian, Joanne Metal, Tammy Sauter and Tatiana Shohov
Circulation: Iyatunde Abdullah, Sharon Britton, and Cathy DeGregory

Library of Congress Cataloging-in-Publication Data
available upon request

ISBN 1-56072-469-2

6080 Jericho Turnpike, Suite 207
Commack, New York 11725
Tele. 516-499-3103 Fax 516-499-3146
E-Mail: Novascience@earthlink.net

Printed in the United States of America

CONTENTS

PREFACE

An international workshop on " Recent Development in Applied Statistics" was held on August 21-23,1996 at Brawijawa University, Malang, Indonesia to honor Professor Munir Ahmad for his contribution to Statistical Sciences. Professor Dr. H.M. Hasyim Baisoeni, Rector, Brawijawa University, Malang, Indonesia inaugurated the workshop. In all, eleven invited speakers covering various topics of statistics participated. There were few contributed papers. Delegates from various Indonesian Universities, Malaysia, Pakistan, India, Iran, Canada, USA and Australia attended the workshop. In this volume, I have included ten papers which were selected of those submitted for publication. I hope that this volume will be of great value to researchers, practicing statisticians and advanced undergraduate or graduate students of statistics.

I would like to express my thanks to the authors who participated at the workshop. My special thank goes to the contributors for their timely contributions and support. I am thankful to Dr. Waego Hadi Nugroho, Dr. Bernadetha Mitakda, and the faculty of the Department of Mathematics, Brawijaya University for their help in hosting the workshop. Thanks are also due to ISOSS and Dr. Khalid Hamid Sheikh, Director ISESCO, Rabat, Morocco. Finally, I would like to offer my thanks to Nova Science Publishers for accepting the task of publishing this volume.

M. Ahsanullah

May, 1997

CONTRIBUTORS

S.E.Ahmed, University of Regina, Regina, Saskatchewan, Canada

M. Ahsanullah, Rider University, Lawrenceville, New Jersey, USA

D.S. Bhoj, Rutgers University, Camden, New Jersey, USA

M. Hanif, King Faisal University,Dammam, Saudi Arabia

H.Hasan, University of Science Malaysia, Penang, Malaysia

S.N.U.A. Kirmani, University of Nothern Iowa, Cedar Falls, Iowa, USA

Dr.Marjono, Brawijaya Univerity, Malang, Indonesia

B. Mitakda, Brawijaya Univerity, Malang, Indonesia

I. Rahimov, University of Science Malaysia, Penang, Malaysia

P.K. Sen, University of North Carolina, Chapel Hill, North Carolina,USA

Bikas K. Sinha, Indian Statistical Institute, Calcutta, West Bengal, India

Bimal K. Sinha, University of Maryland Baltimore County, Baltimore, Maryland, USA

L.A. Sochono, Brawijaya Univerity, Malang, Indonesia

A. R. Soltani, Shiraz University, Shiraz, Iran

MUNIR AHMAD

MUNIR AHMAD

Born in Sialkot, Pakistan, Dr. Munir Ahmad received his early education in statistics from the University of Punjab, Lahore, Pakistan. He obtained a post graduate diploma in statistics (1959) from Aberdeen University, United Kingdom and Ph.D. in statistics (1968) from Iowa State University, Ames, Iowa, USA.

He possesses a wide ranging highly fruitful experience extending over 40 years, of teaching statistics to graduate, post graduate and Ph.D. students at the department of statistics of University of Punjab, Lahore (Pakistan), University of Karachi (Pakistan), Michigan Technological University, Houghton (USA), Tripoli University (now El-Fateh University), Tripoli (Libya) and King Fahd University of Petroleum & Minerals, Dhahran (Saudi Arabia).

He is Fellow of Royal Statistical Society, London, England and holds membership of American Statistical Association ,USA, International Statistical Institute, Netherlands, International Institute of Sample Surveys, Pakistan Statistical Association and is also a member of professional international organizations in the area of quality management systems.

Dr. Munir Ahmad single-handedly founded the Islamic Society of Statistical Sciences (ISOSS) in 1988. He has organized international conferences in Pakistan (1988,1994), Malaysia (1990), Morocco (1992) and Indonesia (1996). These conferences have had wide and sign)ficant influence on the education, research and practices of statistics in the member states of the OIC. Not only that, these meetings have led to fraternal ties among statisticians of the Muslim Ummah as well as the rest of the world.

Dr. Ahmad is an internationally reputed scholar of statistics and has authored/coauthored over 100 research papers in leading statistical journals. He is the founding editor of the Pakistan Journal of Statistics. He is also a contributing editor to Current Index to Statistics of the American Statistical association, USA.

Applied Statistical Science, II
ISBN 1-56072-469-2

NONPARAMETRIC PERSPECTIVES IN REGRESSION ANALYSIS

PRANAB KUMAR SEN
University of North Carolina at Chapel Hill

SUMMARY

The past four decades have witnessed a phenomenal growth of research literature on nonparametric statistical inference (theory and methodology) and its applications in diverse fields of specializations. Nevertheless, there is some cloud over the scope of applicability of some of these sophisticated tools in actual statistical practice. A critical appraisal of the relevant perspectives of nonparametrics is presented here with due emphasis on an important field of application, namely, regression analysis.

1. INTRODUCTION

A little over sixty years ago, *nonparametrics* entered the arena of statistical sciences under the disguise of *quick and dirty* methods. Developers of nonparametrics were mostly from sociology, psychometry, economics and allied fields, who advocated the use of such tools to emphasize the need for making less stringent regularity assumptions for broadening the scope of objective conclusions from their qualitative to quantitative observational studies. During the past forty years, the growth of statistical research literature has been most spectacular, focusing novel theory and methodology, as well as contemplating useful applications in various directions. It has been well recognised that in various interdisciplinary scientific investigations, the scope for strictly parametric statistical analysis is somewhat limited due to possible lack of reliability of model assumptions, and nonparametrics fare well in some respects. This feature has been repeatedly elaborated in contemporary texts and research publications, though there seems to be some ambiguity in a clearcut demarcation of the relative merits and demerits of the parametric and nonparametric approaches. More recent advents in this domain include the so called *robust* procedures as well as *semiparametric* models. Therefore, in depicting this composite picture, we need to take into account the complexities arising from all corners.

Among the various wings of statistical sciences, regression models and their statistical analysis schemes have received the utmost attention. This has been primarily due to the fact that in most scientific, socio-economic and clinical studies, the basic theme of regression analysis has emerged as most appealing. In agricultural, physical, laboratory and other experimentation, it is generally contemplated that a causal relationship exists between the input and output variables, though that relation may be distorted to a certain extent due to measurement errors and other chance variations associated with the experimental schemes. The situation is far more complex for biomedical, clinical and environmental studies where experimentation may not be conductable in a precise controlled setup. In observational studies arising in epidemiological investigations the scenario is even more vulnerable to various uncontrollable factors and calls for more complex statistical modeling and analysis schemes.

*AMS *Subject classifications:* 62G99, 62H99, 62J99

†*Key words and phrases:* Asymptotics; bioassays; dose-response regression; efficiency; error-contaminations; fixed-effects models; generalized linear models; hazard regression; information bounds; L-, M- and R-estimators; Least squares estimators; linear models; measurement errors; maximum likelihood estimators; mixed-effects models; model departures; polychotomous response models; random-effects models; robustness; semiparametrics.

The classical one and multisample models provide the genesis of regression analysis. In comparative studies such samples relate to diverse experimental setups, and hence, the modeling of causal relationship is of natural intertest. These two or several sample location models in turn led to the formulation of simple or multiple regression models. General linear models emerged from such regression models. In this context, the input variables (or effects) may be nonstochastic (or fixed), stochastic (or random) or of mixed type. Moreover, in many experimental schemes an ideal model may not be linear, and suitable nonlinear models are often advocated. *Generalized linear models* (GLM) have emerged in this setup as suitable unifiers of possibly nonlinear models in terms of appropriate linear ones, and this is generally achieved through suitable *link functions* or transformations on variates or statistics. Nevertheless, the regularity assumptions pertaining to such GLM's may not always be tenable in a given practical application, and hence, model flexibility and robustness considerations have prompted further developments wherein semiparametrics and nonparametrics play a vital role. In this study we present an appraisal of this broad scenario with due attention to applications in a variety of fields. In this context, robustness and efficiency considerations dominate our deliberation.

2. LIMITATIONS OF PARAMETRICS

To motivate our discussion, we start with the simple location model in the setup of a *location-scale* family of densities. Let $X_1, \ldots, X_n$ be n independent and identically distributed (i.i.d.) random variables (r.v.) having a contnuous density function $f(x;\boldsymbol{\theta})$, $x \in \mathbf{R}$, $\boldsymbol{\theta} \in \boldsymbol{\Theta}$, where

$$f(x,\boldsymbol{\theta}) = \theta_2^{-1} f_o((x-\theta_1)/\theta_2), \quad \theta_1 \in \mathbf{R}, \quad \theta_2 \in \mathbf{R}^+, \tag{2.1}$$

and the form of the density f_o is free form the unknown $\boldsymbol{\theta}$. Notable examples of f_o are the normal, Laplace, logistic and Cauchy densities. For simplicity of presentation, we treat θ_1 as the parameter of interest and θ_2 is treated as a nuisance parameter. An optimal estimator of θ_1 (attaining the Cramér-Rao information bound to its mean square error) exists when f_o belongs to the *exponential family* of densities, but this characterization does not hold for the Laplace, logistic or Cauchy density. On the other hand, the classical *maximum likelihood estimator* (MLE) exits under more general regularity conditions, and at least asymptotically, it has some optimality properties. For the normal density, the MLE agrees with the classical *least square estimator* (LSE), the sample mean $\bar{X}_n$, it is linear in nature, easy to compute, and is an optimal estimator. This isomorphism of MLE and LSE may not generally hold for nonnormal densities. Moreover, for such nonnormal densities, typically, the MLE is nonlinear in nature, and often requires an extensive iteration scheme for its computation. On the other hand, the LSE is easier to compute, but for distributions with heavy tails (for which the reciprocal of the variance is smaller than the Fisher information on θ_1), the LSE may be inefficient relative to the MLE. For example, if f_o is the Laplace density, the LSE is inefficient relative to the MLE (the sample median), while for the Cauchy density, the LSE is even inconsistent but the MLE is asymptotically efficient. But the MLE may be generally nonrobust to possible model departures, and often this is of serious concern. To illustrate this point, we consider the following example. Suppose that we assume that the density f in (2.1) is normal while actually we have mixture of the following form:

$$f^*(x) = (1-\epsilon)\sigma^{-1} f_o((x-\mu)/\sigma) + \epsilon(k\sigma)^{-1} f_o((x-\mu)/k\sigma), \quad 0 < \epsilon < 1, \quad k >> 1, \tag{2.2}$$

where f_o is the standard normal density. This is the so called *error contamination* model where we allow ϵ to be small, and large values of k indicates heavy tails for distribution of the contaminating variable. For the assumed normal density, the optimal estimator (MLE/LSE) of μ is the sample mean $\bar{X}_n$, and under the contamination model, its variance is

$$n\text{var}\{\bar{X}_n\} = \sigma^{*2} = (1-\epsilon)\sigma^2 + \epsilon k^2 \sigma^2. \tag{2.3}$$

The sample median $\tilde{X}_n$ is a relatively less efficient estimator, and its (asymptotic) mean square error under the contamination model is given by

(2.4) $$\lim_{n\to\infty} n\text{var}\{\tilde{X}_n\} = \frac{\pi\sigma^2}{2}\{1-\epsilon+k^{-1}\epsilon\}^{-2}$$

so that the *asymptotic relative efficiency* (ARE) of the sample mean with respect to the sample median is equal to

(2.5) $$e(\bar{X},\tilde{X}) = \frac{\pi}{2}\{1-\epsilon(1-k^{-1})\}^2\{1+\epsilon(k^2-1)\}.$$

It is easy to verify that for a fixed $\epsilon(>0)$, the ARE is a maximum at $k=1$ (equal to $\pi/2 = 1.57..$), and it monotonically decreases as k moves away from 1. Thus, for every fixed $\epsilon \in (0,1)$, there exists a $k = k_\epsilon$, such that the ARE of the mean with respect to the median falls below 1 for every $k \geq k_\epsilon$. This reveals the nonrobustness of the optimal estimator (MLE/LSE) for the normal distribution to possible error contamination. Contrary to the sample mean, the mdeian is not affected by variations of the extreme values, and hence, it is much more robust to error contaminations or outliers. This sparked the development of robust estimators (Huber, 1964) wherein the idea of combining efficiency with robustness has been coined effectively, and wew shall refer to that later on.

Let us examine the scenario in the multisample case. Suppose that $X_{i1},\ldots,X_{in_i}$ be n_i i.i.d.r.v.'s drawn from a distribution $F_i(x)$, for $i=1,\ldots,k(\geq 2)$, where all the k samples are assume to be independent. In a multisample location model, it is assumed that

(2.6) $$F_i(x) = F_o(x-\theta_i), \quad i=1,\ldots,k,$$

where the θ_i are unknown real location parameters, and the distribution function (d.f.) F_o admits a density f_o having the same structural form as in the single sample case. Our interest centers in relative comparisons of the k location parameters. The situation with the MLE and LSE is quite comparable to the single same case, but a bit more complicated because of the fact that the number of parameters of interest may be greater than one.

The two sample location model directly leads to the *simple regression* model:

(2.7) $$X_i = \alpha + \beta t_i + e_i, \quad i=1,\ldots,n,$$

where the t_i are known regression constants, the intercept parameter α and the regression parameter β are unknown, and $e_1,\ldots,e_n$ are i.i.d.r.v.'s with a d.f. $F_o(e)$. In the same vein, the multisample location model provides the genesis of univariate *linear models*. In that setup, we have

(2.8) $$X_i = \boldsymbol{\beta}'\mathbf{t}_i + e_i, \quad i=1,\ldots,n,$$

where the $\mathbf{t}_i$ are known regression vectors (not all equal), $\boldsymbol{\beta}$ is an unknown regression parameter vector, and the errors e_i are i.i.d.r.v.'s. Though the X_i are not necessarily i.d., the errors are i.i.d., and this feature dominates the statistical analysis of linear models. Conventionally it is assumed that e_i has a normal distribution with 0 mean and a finite (and positive) variance σ^2 (unknown). In this conventional setup the optimal estimator of $\boldsymbol{\beta}$ is the MLE/LSE, and is given by

(2.9) $$\hat{\boldsymbol{\beta}}_n = (\sum_{i=1}^n \mathbf{t}_i\mathbf{t}_i')^{-1}\mathbf{t}_iX_i = \boldsymbol{\beta} + (\sum_{i=1}^n \mathbf{t}_i\mathbf{t}_i')^{-1}\mathbf{t}_ie_i.$$

Thus the MLE/LSE for this normal linear model is linear in the errors e_i, and therefore the wealth of linear statistical inference tools can be used conveniently to draw statistical conclusions on $\boldsymbol{\beta}$. As in the location model, here also, the LSE/MLE exhbits nonrobustness to model departures ranging from possible nonnormality of the errors to other structural assumptions that are usually made in linear statistical inference. We shall discuss these aspects in greater detail here.

In many statistical studies, it may be difficult to contemplate directly a linear model. In such a case, often suitable transformations are advocated for the input and/or output variables, so that a linear model can then be adopted on these transformed variables. Alternatively, some nonlinear models are used to avoid such transformations to a certain extent. As a a simple illustration consider the following model

$$X_i = \alpha e^{-\beta t_i} + e_i, \quad i = 1, \dots, n, \tag{2.10}$$

where the t_i are known regression constants, while as in (2.7), α, β are unknown parameters. In this setup, often, the X_i are nonnegative r.v.'s, so that the assumption of the e_i having a normal distribution remains to be examined carefully. It is not uncommon to prescribe the transformation: $Y_i = \log X_i$, $i = 1, \dots, n$, and consider the model

$$Y_i = \alpha_o + \beta_o t_i + e_i^*, \quad i = 1, \dots, n, \tag{2.11}$$

where $\alpha_o = \log\alpha, \beta_o = -\beta$, and it is tacitly assumed that the errors e_i^* are i.i.d.r.v.'s having a normal distribution with zero mean and a finite, positive variance σ^{*2}. Looking at the last two equations, we are somewhat confused: If the original e_i were normally distributed, could we have taken the transformed e_i^* to be normally or even log-normally distributed? Moreover the variance of the e_i may depend on the level t_i, so that the classical *homoskedasticity* condition may not hold. A similar situation may arise for the error components e_i^*. In many situations, a transformation is primarily chosen to linearize the regression relation, but assuming that normality pertains to such transformed models may not meet the light of reality in practice. As a result, robustness and efficiency considerations may not favor the use of MLE/LSE for such assumed normal error models. Though in some cases, based on theoretical considerations, a specific nonlinear regression relation can be postulated, in a majority of cases arising in biomedical and clinical studies, there may not be enough information to advocate a specific form, so that model departures may also arise due to possibly inaccurate assumed form. We shall discuss this aspect in a later section.

There has been some attempts to unify such parametric models exploiting the linearity and incorporating some nonnormal distributions in a relatively more general framework of GLM's, and the classical *exponential family of densities* provides the genesis of GLM's. We refer to McCullagh and Nelder (1989) for an excellent treatise of GLM's. Consider a density of the form

$$f(y, \theta, \phi) = c(y, \phi)\exp\{(y\theta - b(\theta))/a(\phi)\}, \tag{2.12}$$

where θ is the parameter of interest, $\phi(> 0)$ is a nuisance scale parameter, and $a(.)$, $b(.)$ and $c(.)$ have known functional forms. Then $E(Y) = b'(\theta) = \mu(\theta)$ and $V(Y) = a(\phi)b''(\theta)$; it is assumed that $b(.)$ is continuously differentiable upto the second order. Suppose that $Y_1, \dots, Y_n$ are independent r.v.'s having densities $f(y, \theta_1, \phi), \dots, f(y, \theta_n, \phi)$ respectively. Suppose further that there exists a monotone, differentiable function $g(.)$, termed the *link function*, such that

$$\mathbf{G}_n = (g(\mu(\theta_1)), \dots, g(\mu(\theta_n)))' = \mathbf{X}_n\boldsymbol{\beta}, \tag{2.13}$$

where $\boldsymbol{\beta}$ is an unknown parameter ($q-$vector), and $\mathbf{X}_n$ is a known design matrix of order $n \times q$. If we write $\mathbf{X}_n' = (\mathbf{x}_1, \dots, \mathbf{x}_n)$ and $h(\mathbf{x}_i'\boldsymbol{\beta}) = g^{-1}(\mathbf{x}_i'\boldsymbol{\beta})$, $i = 1, \dots, n$, then the *estimating equations* (EE) for $\boldsymbol{\beta}$ are given by

$$\sum_{i=1}^{n}\{Y_i - h(\mathbf{x}_i'\boldsymbol{\beta})\}\mathbf{x}_i = \mathbf{0}. \tag{2.14}$$

The analogy of the above EE with that of the LSE is quite apparent, and hence, the MLE/LSE methodology can be incorporated for such GLM's too. There are, however, some other complications due to the fact that when $g(.)$ is not a canonical link function, $h(.)$ may not be a linear function (of its argument), so that, as in the case of the MLE for a density not belonging to an exponential family, the resulting EE may have to be solved by iterative procedures. Also as in the case of a logistic regression model, the Y_i

may not be elementary r.v.'s, and in such a case, the scale parameters may also involve the unknown θ_i. Moreover, in the context of *quasi-likelihood* functions, the extension of EE to GEE, considered by Wedderburn (1974) and others, relates to such dependence more explicitly. Therefore, the exact methodology for GLM's may need some modifications which are justifiable mostly in the light of asymptotic theory. In this setup, robustness considerations may not favor an adoption of GLM methodolgy in a parametric framework. We refer to Sen (1996c, 1997) for some detailed discussion of alternative approaches to such GEE's and related GLM's.

We discuss now the limitations of the classical linear statistical inference by reference to the so called *one-way layouts* or *completely randomized designs* covering both fixed-effects and mixed-effects models. Consider first the fixed-effects ANOVA (analysis of variance) model:

$$Y_{ij} = \mu + \tau_i + \epsilon_{ij}, \quad j = 1, \ldots, n_i, \ i = 1, \ldots, k(\geq 2), \tag{2.15}$$

where all the $n = n_1 + \cdots + n_k$ observations are assumed to be independent, and in addition to the lineairty of the model or additivity of the effects, the following assumptions are made explicitly:

(i) $\tau_1, \ldots, \tau_k$ are parameters subject to the restraint that $\sum_{i=1}^k \tau_i = 0$.
(ii) Normality of the errors ϵ_{ij}.
(iii) Independence of the errors.
(iv) Homoskedasticity of the errors.

Note that these assumptions ensures the attainment of the Cramér-Rao information bound for the mean square error of the MLE/LSE of $\boldsymbol{\beta}$, and this, in turn, renders optimality properties for linear statistical inference tools. As in the location model, if the error distribution is not normal, one or more of these assumptions may not be tenable, and as a result, optimality properties of standard normal theory ANOVA procedures may nolonger hold. The extent of damage may depend on the nature and degree of departure from the model assumptions, and we shall illustrate these points later on.

To illustrate the situation with a mixed-effects model, we consider next the classiocal one-way ANOCOVA (analysis of covariance) model:

$$Y_{ij} = \mu + \tau_i + \gamma' \mathbf{Z}_{ij} + e_{ij}, \quad j = 1, \ldots, n_i, \ i = 1, \ldots, k, \tag{2.16}$$

where the $\mathbf{Z}_{ij}$ are concomitant variate(s), and the errors e_{ij} satisfy the conditions (ii), (iii) and (iv) considered for the ANOVA model. Thus, here in addition to all the regularity assumptions in the ANOVA model, we make the following assumptions:

(v) The distribution of the $\mathbf{Z}_{ij}$ does not depend on $j(= 1, \ldots, n_i)$; $i(= 1, \ldots, k)$.
(vi) The regression of Y on $\mathbf{Z}$ is linear.
(vii) Independence of the errors and the concomitant variates.
(viii) $\sigma_o^2 = \text{var}(e_{ij}) = \sigma^2(1 - R^2) \leq \sigma^2 = \text{var}(\epsilon_{ij})$, where R^2, the (squared) multiple correlation coefficient of Y on $\mathbf{Z}$, is strictly positive.

(Otherwise there is no gain in information in the ANOCOVA model over the ANOVA model.) Generally it is assumed that $(\mathbf{Z}, e)$ has jointly a multinormal distribution with a dispersion matrix $\boldsymbol{\Gamma}$ which is block diagonal in the sense that $\text{Cov}(\mathbf{Z}, e) = \mathbf{0}$. This strong assumption ensures the last set of conditions stated above.

Possible sources of departures from the model assumptions include:

(i) Nonnormality of the errors: $\sigma_o^2 > \{I(f)\}^{-1}$, so that there is some loss of information.
(ii) Nonnormality of the conditional distribution of Y, given $\mathbf{Z}$, which may not only affect (iv) and (vi) but also (vii) may be vitiated.
(iii) The additivity of effects in the model may not be true.

These departures may be local or global in nature. In many cases, one may like to use suitable transformation on Y and $\mathbf{Z}$ so as to make the model adequately linear. But then it may not be possible to justify all the assumptions (i) through (viii) on the transformed model; particularly, the normality of the transformed error component, its independence on the transformed covariates, and the homoskedasticity condition may not hold. In some cases, even the independence of the Y_{ij} or the $\mathbf{Z}_{ij}$ may be questionable, raising doubts on the validity and robustness of the classical LSE/MLE under such departures from the model assumptions. In the context of transformed models, generally linearity holds at least approximately, and hence, often, one emphasizes on local departures from normality with associated local departures from the homoskedasticity and other related properties. This has led to the development of *robust statistical* inference tools where robustness relates to local departures and (asymptotic) efficiency aspects dominate the scenario. On the other hand, there are other situations where inspite of the linearity of the model, near normality of the errors may be unreasonable to adopt. This feature favors the robustness aspects in a global sense and leads to alternative procedures that are globally robust. In the context of general linear models, Jurečková and Sen (1996) have an up-to-date treatise of a general class of robust statistical procedures with due emphasis on related asymptotics. This treatise clearly reveal the lack of robustness of standard parametric statistical procedures even for small model departures, and we shall refer to this in a later section.

We discuss a third model to emphasize the lack of robustness of standard parametrics in regression analysis. Consider a set of n bivariate observations (X_i, Y_i), $i = 1, \ldots, n$ drawn from a suitable bivariate distrubtion. Then the *regression function* of Y on X is typically defined in a parametric framework as

$$m(x) = E[Y|X = x], x \in \mathcal{X}. \tag{2.17}$$

If the joint distribution of (X, Y) is bivariate normal, we have

$$m(x) = \alpha + \beta x \text{ and } \sigma_x^2 = \text{Var}(Y|X = x) = \sigma^2, \ \forall \ x \in \ \mathcal{X}. \tag{2.18}$$

Sans the bivariate normality, we may not have the following

(a) Linearity of $m(x)$ in $x \in \mathcal{X}$;

(b) Normality of the conditional distribution of Y, given $X = x$, and

(c) Homoskedasticity, i.e., the constancy of σ_x^2.

Therefore the standard parametric analysis based on an assumed bivariate normal distributional model can be quite vulnerable to plausible departures along each of these routes. Nonparametric regression analysis has its genesis in this simple regression model and we shall discuss that later on.

3. PARAMETRICS IN BIOMEDICAL REGRESSION ANALYSIS

In *bioassays, clinical trials, dosimetric* and *bio-environmental* studies, *dose-response regression* abounds in statistical modeling and analysis. There are various aspects of such dose-response regression models and analysis schemes, more attuned to the specific problem at hand, and often these are in contrast with the classical linear models discussed in the preceding section. This basic difference is primarily due to the difference in the experimental setups as well as in the underlying regularity assumptions that are generally made in order to justify adoptions in actual practice. Moreover such structural differences also call for alternative modes of analysis that suit the practical applications more adequately.

A basic difference in such experimental schemes with the conventional ones is the fact that experimentation may generally involve human beings or subhuman primates, and thereby may have much less control over the variation that can arise due to a multitude of causes, some assignable while others not. In conventional experiments, the input and output variables are generally well defined, and often an axiomatic justification can be made to conceive of a suitable parametric form of regression. On the other hand, in this greater domain of bio-medical, biomechanistic and bio-environmental studies, often, the choice of proper input and output variables may be difficult; there may be too many regressors as well as response variables, and it may be necessary to fathom out the regression relation mostly by incorporating extraneous factors and empirical considerations. In clinical setups, for example, different doses of a drug might have been decided on therapeutic considerations. Apart from the main response for the study of which the preparation is prescribed, administrative and other medical considerations may also require the monitoring of possible side-effects and other toxicities, Therefore, we may have a multiple end-point problem. In addition, often, the response variables are not all continuous, and may even be dichotomous or polychotomous in nature. *Quantal* bioassays are notable examples of this type. Refering back to such bioassays, we may have three different situations: (i) The response is certain (eg., death) while the dose required to produce the response is random in nature; this refers to the so called direct assays. (ii) The dose is generally administered in some stipulated quantities while at each dose level the response may be stochastic in nature. If the response is quantitative in nature, we have the so called indirect quantitative assays. (iii) In the same setup as in (ii), if the response is *all or nothing* (i.e., *quantal* in nature), we have the quantal bioassays. In case (i) usually a logarithm or power (i.e., square root or cubic root) transformation is used on the dose variable to induce approximate normality of the distribution of the transformed variable. In case (ii), such a transformation may be needed on both the dose and response variables; this naturally raises the question: How good are these Box-Cox transformations in indirect quantitative assays, and how robust are the standard parametric procedures based on normal distributional assumptions in such a case when departures from the assumed model can not be ruled out? *Generalized linear models* (GLM) have also been advocated in such parametric setups with the objective of validating standard parametric procedures (based on MLE) in an exact setup but not necessarily incorporating the standard linear models. However, as has been discussed in detail in Sen (1996c, 1997), such GLM based parametric procedures may generally be quite nonrobust to even small to moderate departures from model assumptions. As illustrations consider all the three models considered above. First, in a direct dilution assay, to compare the effectiveness of a test preparation and a standard one, it is assumed that the two tolerance distributions, say, $F_A(.)$ and $F_B(.)$, both defined on $\mathbf{R}^+ = (0, \infty)$, are related by the scale factor ρ in the following way:

$$F_A(x) = F_B(\rho x), \ x \geq 0, \ \ \rho > 0, \tag{3.1}$$

where ρ is termed the *relative potency* of the preparation A with respect to B. In this setup, if we assume that both F_A, F_B are normal, they may not have the same variance (unless ρ is equal to 1). Therefore, it may not be appropriate to use the standard (student's) t-statistics to draw statistical conclusions on ρ. For a valid parametric procedure, one may appeal to the classical Fieller theorem, but there are certain hidden problems there, and often, exactness or efficiency properties may have to be compromized to a certain extent. To eliminate this difficulty, one may consider a *dosage* as the log-dose, so that on this dosage variables, if we assume normality of the two distributions, say $F_A^*(.)$ and $F_B^*(.)$, they would differ only by a shift parameter that is equal to $\log\rho$. However, it may be noted that for the original model, the MLE of ρ would have been a function of the sample sufficient statistics, the two sample means and the sample variance. On the other hand, in the log-dose case, these are to be replaced by the corresponding sample geometric means and the variances of the log-dose values in the two samples. Therefore if contrary to the true model relating to normality of F_A, F_B, we assume that F_A^*, F_B^* are normal, optimal estimators for this model will be generally biased and less efficient for the true model. A similar situation arises when F_A^*, F_B^* are normal, but we take for granted that F_A, F_B are normal. Note that it is not possible to have similutaneous normality of both F_A, F_A^*. This creates a dilemma to a practitioner : When to choose normal vs. log-normal models ? In generaL, it may be the case that the true tolerance distribution is neither normal nor log-normal, so that we may even lose the appeal for standard parametric procedures based on either model. From a practical adoption point of view, it is therefore desirable to have statistical conclusions that remain insensitive to such possible model departures, so that *dosage-invariant* statistical inference tools have greater appeal in such cases. Consider next the case of quantal assays. Here for each preparation, there are several dose levles, so that when administered, some of the responses are positive while the rest not. For such binary responses, the probability of response is assumed to depend on the dosage level according to some simple distribution; among various possibilities, the logistic and normal distributions lead respectively to the *logit* and *probit* models. In the logit model, corresponding to a dose level $\mathbf{x}$, the probability of response, $\pi(\mathbf{x})$, is expressed as

$$\pi(\mathbf{x}) = \{1 + e^{\alpha+\beta'\mathbf{x}}\}^{-1}, \tag{3.2}$$

where α and β are unknown parameters. The logits are then defined as

$$\xi(\mathbf{x}) = \log\{\pi(\mathbf{x})/(1-\pi(\mathbf{x}))\} = \alpha + \beta'\mathbf{x}, \tag{3.3}$$

so that there is a linear relationship which can be exploited through the estimates of the $\xi(\mathbf{x})$. On the other hand, in a probit model, we take

$$\pi(\mathbf{x}) = \Phi(\alpha + \beta'\mathbf{x}), \tag{3.4}$$

where $\Phi(.)$ stands for the standard normal distribution function. The probits are defined as

$$\xi(\mathbf{x}) = \Phi^{-1}(\pi(\mathbf{x})), \tag{3.5}$$

which also reveals a linear relationship that can be exploited through the estimates of the probits in the sample. There are various problems associated with either the logit or probit methodology. First, the estimators of the $\xi(\mathbf{x})$ do not have the exact normal or some other simple distribution so as to facilitate the applicability of the LSE/MLE methodology, and there are subtle asymptotic undercurrents which are needed to be taken into account. Secondly, the conditional mean square error of such an estimator, given $\mathbf{x}$, may depend on the unknown $\pi(\mathbf{x})$, so that the usual homoskedasticity condition prevailing in simple linear models may not hold in this case. As a result the estimating equations (based on the modified GLM approach) call for the weighted LSE procedure, and thereby are computationally cumbersome. Finally, in actual practice, it may not be known whether the true model is given by the logit or probit one, so that if the assumed model is different from the true one, the derived estimators of α, β may be biased, inefficient, and may even be inconsistent. Thus, such GLM based parametrics may not be generally robust. This lack of robustness is particularly crucial when the $\pi(\mathbf{x})$ are not all clustered in the central part of (0,1). In many medical, environmental and epidemiological studies, experimentation is made with particularly low doses for which the corresponding $\pi(\mathbf{x})$ are close to the lower extremity (i.e., 0). In such a case, a parametric model may exhibit significant degree of nonrobustness.
In the case of indirect quantitative assays, generally both the (nonstochastic) dose and (stochastic) response variables are quantitative, often, continuous in nature. But the dose response regression may be highly nonlinear, and hence, suitable transformations are made on both the dose and response variates; these are called the *dose metameter* (or dosage) and *response metameter* respectively. With these transformed variables a linear regression model is contemplated for further statistical analysis. Here also even if a correct linear regression model is assessed, the errors may not have exactly the normal or logistic distribution, so that standard linear inference tools can be routinely used. Heteroskedasticity is typically the case with these transformed errors, and as a result, the related estimating equations may be generally cumbersome. Finally, in general, there may not be enough information to choose judiciously the dose and response metameters, and hence, the lack of robustness for such model departures, even by small amount, remains a big issue for such related parametrics.

For all the three models, there are other problems that may affect parametrics quite drastically. First, consider the case where measurement of the exact dose level may not be feasible, and the administered levels are thereby subject to some *measurement errors* or in the discrete case, *misclassifications*. Such errors vitiate the simple probability structure undering the observed events (such as the logit or probit models), and as a result, ignoring them in statistical modeling and analysis may affect seriously the statistical properties of the estimators. A similar case arises when the response variable is also subject to such measurement errors or misclassifications. Secondly, in any experimentation, it is not uncommon to encounter a situation where a majority of the sample observations come from the true distribution, but a small fraction of observations, called the *outliers* come from some other population which may have much heavier tails (or dispersion); this is generally referred to as the *gross error* or *error contamination* model. In the conventional linear model, led by Huber (1973), researchers have studied the effects of small to moderate amount of contamination on the performance characteristics of standrard parametric procedures. They reveal lack of robustness, often to a considerable degree, and some alternative procedures, termed *M-procedures* fare much better in such cases. In the current context, inspite of having approximate linearity of regression, in the absence of the exact normality of the errors and their homskedasticity, the extent of nonrobustness under contamination may be much higher.

In biomedical applications, mostly, multiparameter models are encountered, and often these relate to multivariate observations. The complexities underlying the characterization of optimal statistical procedures in such complex biometric models may make the adoption of standard parametrics less appealing than in the conventional case of single parameter or the family of multiparameter exponential densities. Even in the conventional i.i.d. case, a classical example of a type of parametric paradox is the so called *Stein phenomenon*, and is due to Stein (1956), who considered multinormal distributions of dimension ≥ 3 (a member of the multiparameter exponential family), and showed that the classical MLE of the population mean vector, is not admissible under quadratic error loss functions, and there exit some other estimators, known as the *shrinkage (Stein-rule)* estimators which dominate the MLE in quadratic risk. During the past forty years, a phenomenal growth of research literature on shrinkage estimation has taken place and these developments are no longer confined to multinormal or generalized linear models. Nevertheless, in biomedical applications a more pertinent question is the appropriateness of conventional parametric models, and hence, such optimality properties are to be judged with some caution with due considerations on validity and robustness properties. Coming back to the Stein phenomenon, we may mention that nonparametrics and semiparametrics have also been annexed to this complex in a natural manner. An approach based on the (generalized) *Pitman measaure of closeness* (PMC) of Statistical estimators (Pitman, 1937), elaborated to a greater extent in Keating, Mason and Sen (1993), provides further insights into the Stein paradox, under even more general setups (e.g., covering the bivariate models as well).

Let us move to another important area in biomedical studies, namely, the *clinical trials*. In a simple setup, suppose that there are two groups of subjects, the *placebo* and the *treatment* group, and we intend to compare their responses in order to analyse the treatment effects in an objective manner. Typically the response variable is the *time to failure* which is nonnegative. Moreover, usually there is a time-sequential element in the experimental setup, so that the smallest ovservation is observed first, the second smallest next, and so on, with the largest one occuring at the end. Even if there is no difference in the two response distributions, these ordered variables, representing the *order statistics* of the combined sample from a single distribution, are neither independent nor identically distributed. Moreover along with the primary response variable, there are usually a number of concomitant variates and possibly some other auxiliary ones too. This brings the *induced order statistics* or *concomitants of order statistics* in the picture. These statistics possess some conditional independence property, but they are not generally identically distributed. Further, due to time and cost restraints, an experiment or clinical trial can not be conducted until all the failures occur, and usually the trial is curtailed after a prefixed duration of time, resulting in a truncation (called *Type I cencoring*), or after a prefixed number of respondents are observed (called, *Type II censoring*). In either case, there is some incompleteness in the acquired dataset due to censoring, and this, in turn, leads to some loss of statistical information. More complex forms of censoring, such as the *random censoring*, crop up due withdrawl or dropouts or even due to staggering entry plans. Due to these reasons, the conventional linear models may not be generally appropriate here even if we use the logarithmic or other Box-Cox type transformations on the primary and concomitant variables. Therefore, parametric approaches to such statistical modeling and analysis schemes may not generally be suitable. Both semiparametric and nonparametric approaches have been worked out extensively in this context, and their relative suitability depends on the situation at hand.

The simplest semiparametric approach is due to Cox (1972). He assumed that the conditional *hazard function*, given the concomitant variates, can be expressed as a product of (an arbitrary) baseline hazard function, and a nonnegative function of parametric nature involving the regression on the concomitant variables alone. Cox advocated an exponential regression function so that in terms of the logarithm of the conditional hazard function one has a linear regression with the intercept given by the logarithm of the baseline hazard function. In the original formulation we have the so called *proportional hazard model* (PHM), while the log-transformation leads to the hazard regression model. Inspite of the assumed linearity of regression, statistical analysis schemes are more complex in such semiparametric models, primarily due to the fact that the baseline hazard function being nonparametric in nature needs a more delicate treatment. Cox (1972) essentially proposed a *partial likelihood* approach wherein for the estimation of the hazard regression parameter(s), one can treat the baseline hazard as nuisance, and albeit losing some information due to conditioning, one can arrive at consistent and reasonably efficient estimates. The extent of loss of efficiency depends on the extent of censoring, so that under heavy censoring this procedure becomes quite inefficient. Cox's PHM is essentially built around the classical *log-rank statistics* which is in turn a member of a general class of rank prodecures that lead to the core of nonparametric procedures, discussed in detain in Sen (1981, Ch. 11). Basically, the PHM assumption is not that crucial for nonparametrics, and there are some notable situaions where the PHM may not be appropriate but such nonparametrics work out better; we refer to Sen (1994) for some illustrative examples of this type. Moreover, the partial likelihood approach is quite sensitive to the appropriateness of the PHM, and as a result, the Cox analysis may not be generally robust for departures from the basic PHM. On this count nonparametrics fare better. Even for the PHM, there are various modifications to suit more complex situations caused by multiple end-points, surrogate end-points, misclassified models, measurement errors, etc., and in such a case, there may be a greater appeal for the nonparametric approach which can accommodate these complexities in a more robust manner. Consider for example a multiple endpoint clinical trial involving possibly some surrogate endpoint. Typically, we may have a *mixed response* model wherein some of the variables are continuous while the others are either dichotomous or polychotomous, or at best have some discrete distribution. This may require the incorporation of suitable transformations on the continuous variable to induce (multi-)normality, and logit/probit type transformations on the binary ones. Often, we may not have any precise knowledge of such transformations, and they are mostly chosen on the ground of their mathematical tractability! This may of course result in a lack of robustness property, particularly, when the number of observations is not so large.

Repeated measurement designs arise frequently in follow-up studies as well as in biomedical and clinical experiments. Typically the same group of subjects are either followed through over time or they have repeated measurements or responses at various time points. Naturally, for the same subject, the responses at various time points are generally stochastically dependent. In the simplest setup, a multinormal model is incorporated wherein the entire spectrum is characterized in terms of the mean vector and dispersion matrix. Thus, a finite dimensional parameter space pertains to the model, and granted the inherent linearity and homoscedasticity of multinormal distributions, linear statistical inference becomes largely adoptable. Even one may like to impose specific type of dependence (or correlation) structures to simplify further the statistical analysis schemes. On the other hand, the scope for departures from such an assumed model is far more in such a case than in conventional linear models. A *Markov Chain* model is also adopted for such repeated measurement designs. Though it induces some flexibility in the distributional regularity assumptions, in terms of robustness perspectives, such models may not be very attractive. For these reasons, more and more semiparametric and nonparametric methods having better robustness perspectives are being considered in the literature. We shall refer to some of these in the subsequent sections.

4. WHITHER SEMIPARAMETRICS ?

In prescribing appropriate statistical models and analysis schemes in specific applications in a greater domain of applied and experimental (including biomedical, clinical and health) sciences, there are two basic prerequisites :

(i) Whether the given experimental setup justifies the adoption of a particular parametric model (structure) for the observable responses ?

and (ii) how good are the distributional assumptions in the given context?

Largely because of computational ease and simplicity of formulations, specific parametric models have been incorporated in statistical analysis schemes. For example, linear models based on normally distributed error components have been most commonly adopted in regression analysis, while in hazard regression studies (and in reliability theory) the exponential and Weibull distributions are generally incorporated in statistical modeling and analysis. Yet in actual practice there may not be sufficient evidence that such simple models are tenable in the specific cases. Use of transformations of variables and statistics has also been popularized to enhance the scope of applicability of such standard parametric statistical procedures, albeit there may not be any guarantee that normality or exponentiality of the transformed errors would be achieved along with the primary goal of achieving linearity of regression or other structural relations. Semiparametric procedures have mainly been motivated to eliminate some of the major drawbacks of classical parametric procedures, and validity and robustness considerations dominate this scenario, and such considerations differ from the usage of nonparametric procedures to a certain extent. In semiparametric formulations, part of the model has largely a parametric flavor while the complementaty part is mainly nonparametric in nature. We introduce a semiparametric model through some simple illustrative models.

(i) *Semiparametric linear model.* Consider the univariate linear model, presented in (2.8), sans the normality of the errors e_i. Instead, assume that the errors are i.i.d.r.v.'s with a continuous (but unknown) distribution $F(x)$, defined on the real line R. Even the symmetry of F may not be very crucial in this context. Thus, we have a parametric (linear) regression relation but a nonparametric error distribution, and hence, we term it a semiparametric linear model. In this broad setup, we may point out a discriminating point of view concerning the domain of F. Often we have a reasonable degree of confidence on a small subclass $\mathcal{F}_o \in \mathcal{F}$, the entire class of all continuous F, so that we restrict ourselves to the domain $F \in \mathcal{F}_o$. For example, $\mathcal{F}_o$ may stand for a very local class of distributions which differ from a specified normal distribution by small amount (say, by error-contamination or in a more sophisticated topolgy). This particular scenario favors the incorporation of the so called M-procedures based on robust and locally efficient M-statistics, considered by Huber (1964, 1973) and others. On the other hand, if the true F does not belong to this small subclass, the robustness and efficiency properties may have to be compromised to a certain extent. In this context, if $R-$procedures, based on appropriate rank statistics, are considered, they remain valid for the entire domain $\mathcal{F}$, though may not be fully efficient over the subclass $\mathcal{F}_o$. Side by side, one may also consider $L-$procedures based on linear combinations of functions of order statistics, and these have good robustness properties mingled with good efficiency perspectives too. Such procedures are therefore termed globally robust ones. Moreover, for these alternative approaches in semiparametric linear models, if we confine ourselves to the class of absolutely continuous density functions having finite *Fisher information* with respect to location, then an *adaptive procedure* based on data based estimated score function can be formulated in such a way that it is asymptotically optimal within this class of densities. For an up-to-date treatise of robust statistical procedures in linear models, we may refer to Jurečková and Sen (1996) where $L-, M-$, and $R-$ procedures along with their siblings have been considered and their interrelations have been explored in a systematic and unified manner. Puri and Sen (1985) also contains an earlier account of rank based statistical infertence for general linear models covering univariate as well as multivariate response variables.

(ii) *Semiparametric hazard regression models.* In survival analysis and reliability studies, life distributions are usually compared in the light of their associated hazard functions or failure rates. In this setup, often, there are pertinent concomitant variates, so that the dependence of the hazard rate on such covariates is of genuine interest. Since life distributions have support $\mathcal{R}^+$ (or the multivariate positive orthant), the very assumption of (multi-)normality may not always be very reasonable. Even working with the logarith transformation may not universally lead to a Gaussian model. On the other hand, the Weibull family (including the simple exponential density) may not possess all the desirable properties one would like to have in fruitful statistical modeling and analysis. For example, the simple exponential model is characterized by a constant hazard rate while for the Weibull density, this is given by x^γ, $x \in \mathcal{R}^+$, $\gamma > 0$. In practice this simple setup may not match the sample counterpart adequately. For example, if the life distribution is a mixture of two simple exponential or Weibull distributions, the associated hazard rate has a more complex form and is not so much amenable for simple statistical analysis. Moreover, if such statistical setups, it is not uncommon to have withdrawls or dropouts, resulting in censoring or incompletness of certain kind. This leads to more complex models where the conventional statistical methodology may not work out well. Thus, standard parametric models are generally not so robust in actual applications.

The first spark of a semiparametric approach is due to Cox (1972) who introduced the notion of proportional hazards, treating the baseline hazard rate as arbitrary while formulating the hazard regression in a simple parametric form. He conveyed the idea of incorporating *partial likelihood functions* in statistical analysis (albeit in a somewhat heuristic manner), and during the past 25 years a steady flow of research has led to a solid foundation of semiparametrics in survival and reliability analysis. This approach is heavily dominated by martingale theory adapted to multivariate counting processes having some sort of multiplicative intensity functions. Andersen et al. (1993) contains a systematic account of these developments. In this context it is quite appropriate to point out the the drawbacks this approach may have in applications in practice; of course, more work has been in progress to eliminate some of these drawbacks, though at the cost of more complexity of statistical modeling and analysis.

We have already noticed earlier that the Cox PHM may not be very appropriate when the survival functions cross over each other. Therefore, to incorporate the counting process methodolgy in real applications, it is necessary to check the validity of the multiplicative intensity functions assumption. Secondly, in many situations, the covariates are themselves *time-dependent* so that the constancy of the (hazard) regression relationship over time may not be generally true. To eliminate this drawback, an alternative formulation (cf. Murphy and Sen, 1991) may be to consider *time-dependent regression coefficients.* In this generalization, the emphasis on the PHM is lost to a greater extent. On the top of that one would have then a functional (regression) parameter space, in addition to having a functional space for the unknown baseline hazard function. As a result, *smoothing techniques* incorporating *binning* of the time-parameter are generally employed to achieve consistency of the estimators. This in turn requires a comparatively larger sample size and entails some bias functionals that are of the same order of smallness as the anticipated slower rates of convergence of the estimators. Thirdly, measurement errors and misclassification of states with the covariates may also invalidate the multiplicative intensity assumption to a certain extent. Finally, Multiple and surrogate end points may create additional complications in statistical modeling and analysis of such counting processes based data sets. There seems to be ample room for relaxing the basic regularity assumption needed for the concept of a surrogate end point, and this would probably call for more complex methodological supports for such statistical analysis schemes. The situation with multiple end points is far more complex not only due to the copmplexities and relative importance of these end points but also due to their interdependence in clinical as well as statistical senses. Pedroso de Lima and Sen (1997) have exhibited the intricate connections between some parametric models and their semiparametric counterparts based on *matrix-valued counting processes.* This area is full of promise of more advanced reserch to match the level of complexities in practical applications. These discussions clearly point out the limitations of semiparametric approaches. In the next section, we present some nonparametric counterparts and compare the two approaches.

5. NONPARAMETRICS: GENERAL RATIONALITY

Compared to the parametrics, in semiparametrics, we have at least a nonparametric component that renders some model flexibility. Yet in practice, often, the postulated parametric component may not match well with the true model. Nonparametrics have the main advantage of this greater model flexibility, though this may sometime demand a comparatively larger sample size to validate the usual asymptotics that accompanies nonparametric statistical analysis. To illustrate this point, let us go back to the multisample model considered in (2.6). Opposed to the particular location model treated there, we may consider a broader class of alternatives that the distributions $F_1, \ldots, F_k$ satisfy a *stochastic ordering* in the sense that there exists a permutation of $(1, \ldots, k)$, say $(i_1, \ldots, i_k)$, such that

$$F_{i_1}(x) \geq F_{i_2}(x) \geq \cdots \geq F_{i_k}(x), \ \forall x, \tag{5.1}$$

with strict inequality in at least one of the places for a subset of x values having a positive measure. The shift in location model in (2.6) is a special case of (5.1) when the θ_i are not all equal. Another important subclass of distributions satisfying the model (5.1) is the so called Lehmann (1953) alternative models, where we set

$$\bar{F}_i(x) = 1 - F_i(x) = \{\bar{F}_j(x)\}^{c_i/c_j}, \ \forall\ i \neq j = 1, \ldots, k, \tag{5.2}$$

and the c_i are positive constants. Basically, for the multisample model, against such stochastic ordering, the ranks of the sample observations, being invariant under any strictly monotone transformation on the sample observation, constitute the *maximal invariants*. For this reason statistical inference procedures based on ranks have some distinct advantages over other forms of statistics. Let us consider the univariate linear model outlined in (2.8). In this context we may consider a natural generalization of (5.2) as

$$\bar{F}_i(x) = \{\bar{F}\}^{g(\boldsymbol{\beta}'\mathbf{t}_i)}, \ i = 1, \ldots n, \tag{5.3}$$

where the $\mathbf{t}_i$ are known (regression) vectors, $\boldsymbol{\beta}$ is an unknown regression parameter vector, and $g(\cdot)$ is a nonnegative function of known form. It is of interest to note that the Cox (1972) proportional hazard model considered earlier corresponds to the special case

$$g(\boldsymbol{\beta}'\mathbf{t}_i) = e^{\boldsymbol{\beta}'\mathbf{t}_i}, \ \forall\ i \geq 1. \tag{5.4}$$

Likewise, the semiparametric linear model treated in the preceding section corresponds to the special case where $F_i(x) = F(x - \boldsymbol{\beta}'\mathbf{t}_i)$, $i = 1, \ldots, n$, though in the current nonparametric setup, we may allow the d.f. F to be quite arbitrary. Therefore the emphasis of normality on F in the parametrics or even the near normality in the semiparametrics can be totally dispensed with in nonparametrics. In view of this model flexibility, we may have a nonparametric look into the ANOVA models treated in Section 2.

In the one-way layout model in (2.15), we may simply state the null hypothesis in terms of the homogeneity of the d.f.'s $F_1, \ldots, F_p$ and postulate the alternatives in terms of some stochastic ordering of these d.f.'s. In this formulation, addtivity of the treatment effects, homoskedasticity and normality of the errors, and even the finiteness of the error variances can be replaced by much less restrictive regularity assumptions. On the other hand if the basic regularity assumptions in parametric or semiparametric models hold, such nonparametric procedures remain quite efficient; it is possible to choose some adaptive nonparametric procedures that are asymptotically fully efficient for such (semi-)parametric formulations. A very similar situation arises in the so called two-way layout models. As a matter of fact, the conventional *intrablock ranking* procedures may not even need the mensuration of the actual responses; their relative ranks provide all the needed information for the computation of statistics based on the *method of n-raknings*. Analysis of covariance (ANOCOVA) models have also been adopted to nonparametrics, resulting in greater model flexibility, and thereby greater scope of adaptability in diverse setups where the regularity assumptions underlying (semi-)parametrics may not be that appropriate. For detailed study of the properties of such procedures, we refer to Puri and Sen (1985) and Jurečková and Sen (1996). Rank based procedures are also very appropriate in semiparametrics. For example, consider the usual two-way layout model with possibly random block-effects and fixed treatment effects, and assume that the linearity of the treatment effects holds but the stochastic components are not necessarily normally distributed. In this situation, it is possible to use suitable *alignment* procedures that eliminate the stochastic block effects at the cost of a structured dependence pattern for the within block aligned observations. This permits the use of conventional rank statistics (termed aligned rank statistics), having suitable nonparametric structures under some conditional setups, and asymptotically they lead to efficient inference procedures. We refer to Chapter 7 of Puri and Sen (1971) for a treatise of aligned rank tests in two-way layouts, covering the conventional randomized block designs along with some incomplete block designs as well. The main feature of such R-tests and R-estimates is their robustness to possible departures from normality, beyond the usual infinitesimal neighborhood as is usually contemplated in M-tests and M-estimators. There is, however, a discouraging feature of such R-estimators for general linear models. For a simple regression model, for monotone score generating functions, aligned rank statistics behave as monotone functions of the regression parameter (and are translation-invariant), though this may not be usually linear in nature. Asymptotic uniform linearity of rank statistics in regression parameters, presented in greater depth in Jurečková and Sen (1996), provides nice asymptotic resolutions for the actual computation of R-estimators. In a general linear model, the monotone nature of aligned rank statistics in the regression parameters may be difficult to verify, thogh the asymptotic uniform linearity result in a shrinking neighborhood of the true parameter point remains in tact. This makes it often difficult to solve for the R-estimators in closed forms, and generally, an iterative procedure, based on suitable preliminary estimators and incorporating the above linearity result, is employed to facilitate actual computations. These complexities may mount up for large dimensional parameter spaces. Recent years have witnessed phenomenal growth of research literature on alternative estimators that fare better from the computational point of view, and, at the sametime, share good efficiency properties of the classical rank procedures. Among these, special mention may be made of the *regression quantiles*, developed by Koenker and Bassett (1978), and *regression rank scores estimators*, developed by Gutenbrunner and Jurečková (1992). For an up-to-date account of these developments, along with some general asymptotic equivalence results, we again refer to Chapter 6 of Jurečková and Sen (1996).

Let's turn our interest to nonparametric regression, and to motivate this, we refer to the model considered in (2.17) - (2.18), where various regularity assumptions are listed systematically. In actual practice, it is quite likely that one or more of these basic conditions may not be tenable. The damage that may result from such departures from the postulated normal model may be quite extensive, depending on the source(s) of the departure(s). Therefore, from robustness consideration it is quite important to define a nonparametric regression as in (2.17), and allow it to be rather arbitrary, though subject to some comparatively milder regularity assumptions. The first and foremost concern is the very linearity of $m(\mathbf{x})$ postulated for the normal case. Moreover, being the conditional expectation, $m(\mathbf{x})$ may not only inherit the technical problems of interpretations of a conditional expectation, but also possesses the lack of robustness property of linear functionals with unbounded influence functions. For this reason, other *conditional statistical functionals* have been proposed by a host of researchers. Among these, special mention may be made of the *conditional quantile function*, i.e., the quantile of the conditional distrubution, given $\mathbf{x}$, and *Hadamard differentiable* conditional functionals. These developments have been extensively reviewed in a general multivariate setup by Sen (1993), and pertinent references were all cited there. Basically, there is a need to incorporate suitable *smoothing techniques* to reduce the magnitude of bias that may crop up due to the intricacies involved in the definition of the sample or empirical conditional distribution function. For this purpose, either the well known *kernel* method of smoothing is incorporated, or one may use the *nearest neighborhood* methodology to accomplish the same job. In either case, a binning is adopted with the convention of a bandwidth converging to 0 with $n \to \infty$ in such a way that there is a balance between the asymptotic bias term and the related asymptotic mean square error. The principal shortcoming of such smoothing procedures is that such estimators have slower rates of convergence compared to the conventional estimators in parametric models with a fixed dimensional parameter space. On the other hand, in terms of validity and robustness considerations, smoothing techniques score much higher than their semiparametric or parametric counterparts.

Nonparametric regression analysis has also been adopted in mixed-effects models where part of the effects are nonstochastic and the rest stochastic in nature; we refer to the model in (2.16) as an important illustration. For that ANOCOVA model, the treatment effects were assumed to be fixed while the concomitant variables being stochastic pertain to random effects. For the treatment effects it may not be that unreasonable to assume additivity, but for reasons well known, it may not be very wise to impose the regularity assumptions concerning the linear regression of the response variable on the concomitant ones and the fine structure on the error variables. Thus, it may be more effective to consider a model where linearity holds only to a limited extent while nonparametric regression relation pertains to the complementary part. In this way the usual homoskedasticity condition of the error distribution at various levels of the concomitant variates, their normality or other conditions may not be that crucial. Alignment principle works out well in this context. As by definition the concomitant variates are i.i.d., the marginal distributions of the response variables for various combinations of the fixed effects conform to a semiparametric linear model. Therefore, using either semiparametric techniques or classical nonparametric ones, these fixed-effects parameters can be estimated satisfying the conventional $\sqrt{n}$-consistency property; these estimators, however, are not generally fully efficient relative to the full mixed-effects model. In the next step, using these marginal estimators of the fixed-effects parameters, residuals or aligned observations are obtained in the usual manner; these aligned observations are, however, not generally i.i.d.r.v.'s. Since the rate of convergence for the nonparametric regression function is typically slower than $O(n^{-1/2})$, and the perturbations due to the alignment procedure are $O_p(n^{-1/2})$, the classical nonparametric regression methodology applicable to i.i.d. errors remains adoptable for such aligned observations as well (see, Sen (1996a,b)). Therefore we arrive at nonparametric estimators of the regression function of the response variable on the concomitant variates, and these have the same rate of convergence as the usual ones based on the true errors. In the next step, we may stratify the set of observations on the basis of the concomitant variates into a (moderately) large number of grids, and within each grid, we use the conventional marginal model to estimate the fixed-effects parameters. These estimates from the different grids are then pooled into a combined estimator of the fixed-effects parameters by the usual weighted least squares methodology with the weights determined from the estimated dispersion matrices of the individual within-grid estimators. This way, we arrive at a more efficient estimator of the fixed-effects parameters. For details, we refer to Sen (1996a,b). Such improved estimators of the fixed-effects parameters are then incorporated in the computation of aligned observations or residuals, and the nonparametric regression technique is reemployed to yield the final step estimator of the regression function of the primary response variable on the concomitant ones. This procedure amends itself well to two-way layouts as well as to multivariate models.

REFERENCES

Anderson, P. K., Borgan, O., Gill, R. D., and Keiding, N. (1993). *Statistical Models Based on Counting Processes*, Springer-Verlag, New York.

Cox, D. R. (1972). Regression models and life tables (with discussion). *J. Roy. Statist. Soc.* **B34**, 187-220.

Finney, D. J. (1978). *Statistical Methods in Biological Assay* 3rd Ed., Griffin, London

Gutenbrunner, C., and Jurečková, J. (1992). Regression rank scores and regression quantiles. *Ann. Statist.* **20**, 305-330.

Hampel, F. R., Rousseeuw, P. J., Ronchetti, E., and Stahel, W. (1986). *Robust Statistics - The Approach Based on Influence Functions.* John Wiley, New York.

Huber, P. J. (1964). Robust estimator of a location parameter. *Ann. Math. Statist.* **35**, 73-101.

Huber, P. J. (1973). Robust regression: Asymptotivs, conjectures and Monte Carlo. *Ann. Statist.* **1**, 799-821.

Jurečková, J., and Sen, P. K. (1996). *Robust Statistical Procedures: Asymptotics and Interrelations.* John Wiley, New York.

Keating, J. P., Mason, R. L., and Sen, P. K. (1993). *Pitman's Measure of Closeness: A Comparison of Statistical Estimators.* SIAM, Philadelphia.

Koenker, R., and Bassett, G. (1978). Regression quantiles. *Econometrika* **46**, 33-50.

Lehmann, E. L. (1953). The power of rank tests. *Ann. Math. Statist.* **24**, 23-43.

McCullagh, P., and Nelder, J. A. (1989). *Generalized Linear Models*, 2nd Ed., Chapman-Hall, London.

Murphy, S. A., and Sen, P. K. (1991). Time-dependent coefficients in a Cox-type regression model. *Stochas. Proc. Appl.* **39**, 153-180.

Pedroso de Lima, A. C., and Sen, P. K. (1997). A matrix-valued counting process with first-order interactive intensities. *Ann. Appl. Probab.* **12**, in press.

Pitman, E. J. G. (1937). The closest estimates of statistical parameters. *Proc. Cambridge Phil. Soc.* **33**, 212-222.

Puri, M. L., and Sen, P. K. (1971). *Nonparametric Methods in Multivariate Analysis.* John Wiley, New York.

Puri, M. L., and Sen, P. K. (1985). *Nonparametric Methods in General Linear Models.* John Wiley, New York.

Sen, P. K. (1981). *Sequential Nonparametrics: Invariance Principles and Statistical Inference.* John Wiley, New York.

Sen, P. K. (1993). Perspectives in multivariate nonparametrics: Conditional functionals and ANOCOVA models. *Sankhyā, Ser A,* **55**, 516-532.

Sen, P. K. (1994). Some change-point problems in survival analysis: Relevance of nonparametrics in applications. *J. Appl. Statist. Sci.* **1**, 425-444.

Sen, P. K. (1995). Censoring in theory and practice : Statistical perspectives and controversies. *Analysis of Censored Data* (eds. H. Koul and J. V. Deshpande), IMS Lect. Notes Mon. Ser. No. 27, Hayword, Calif. pp. 177-192.

Sen, P. K. (1996a). Robust and nonparametric methods in linear models with mixed-effects. *Tetra Mount Math. Publ.* **7**, pp. 331-343.

Sen, P. K. (1996b). Regression rank scores estimation in ANOCOVA. *Ann. Statist.* **24**,

Sen, P. K. (1996c). Genelaized linear models in biomedical applications. *Applied Statistical Analysis* (eds. M. Ahsanullah and D. Bhoj.), Nova Press, NJ., pp. 1-22.

Sen, P. K. (1997). A critical appraisal of generalized linear models in biostatistical analysis. *J. Appl. Statist. Sci.* **4**, in press.

Wedderburn, R. W. M. (1974). Quasi-likelihood function, generalized linear models and the Gauss-Newton method. *Biometrika*, **61**, 439-447.

Applied Statistical Science, II
ISBN 1-56072-469-2

RECORD VALUES OF UNIVARIATE DISTRIBUTION

M. Ahsanullah
Department of Management Sciences
Rider University Lawrenceville, NJ 08648-3099, USA

1. Introduction

Suppose that $X_1, X_2, ...$ is a sequence of independent and identically distributed random variables with distribution function F(x). Let $Y_n = \max(\min)\{X_1, X_2, ..., X_n\}$ for $n \geq 1$. We say X_j is an upper(lower) record value of $\{X_n, \ n \geq 1\}$, if $Y_j > Y_{j-1}, j > 1$. By definition X_1 is an upper as well as a lower record value. One can transform the upper records to lower records by replacing the original sequence of $\{X_j\}$ by $\{-X_j, j \geq 1\}$ or (if $P(X_i > 0)=1$ for all i) by $\{1/X_i, i \geq 1\}$; the lower record values of this sequence will correspond to the upper record values of the original sequence. Unless mentioned otherwise we will call the upper record values as record values. The indices at which the record values occur are given by the record times $\{U(n)\}$, $n \geq 0$, where $U(n) = \min\{j|j>U(n-1), X_j > X_{U(n-1)}, n>1\}$ and $U(1) = 1$. The record times of the sequence $\{X_n \ n \geq 1\}$ are the same as those for the sequence $\{F(X_n), \geq 1\}$. Since F(X) has an uniform distribution, it follows that the distribution of $X_{U(n)}$, $n \geq 1$ does not depend on F. We will denote L(n) as the indices where the lower record values occur. By our assumption U(1) = L(1) = 1. The distribution of L(n) also does not depend on F.

For a given set of n observations, let $X_{1,n} \leq X_{2,n} \leq \dots X_{n,n}$ be the associated order statistics.Suppose that $P\{a_n (X_{n,n} - b_n) \leq x\} \xrightarrow{d} G(x) \ as \, n \to \infty$ · For necessary and sufficient conditions about this convergence for various distributions see Galambos (1987). It is well known (see Leadbetter et al. ,1983,p.33) that

$$P(a_n (X_{n-m,n} - b_n) \leq x) \xrightarrow{d} G(x) \sum_{s=0}^{m} \frac{\{-\ln G(x)\}^s}{s!}$$

It can be shown that the right side of the above equation is the distribution function of the m+1th lower record value. Properties of record values of i.i.d. rvs have been extensively studied in literature, for example, see Ahsanullah (1988), Arnold, Balakrishnan and Nagaraja(1992) Nagaraja (1988) and Nevzerov (1987) for recent reviews.

Many properties of the record value sequence can be expressed in terms of the function R(x), where $R(x) = -\ln \bar{F}(x)$, $0 < \bar{F}(x) < 1$ and $\bar{F}(x) = 1 - F(x)$. Here 'ln' is used for the natural logarithm.If we define $F_n(x)$ as the distribution function of $X_{U(n)}$ for $n \geq 1$, then we have

$$F_1(x) = P[X_{U(1)} \leq x] = F(x) \tag{1.1}$$

$$\begin{aligned} F_2(x) &= P[X_{U(2)} \leq x] \\ &= \int_{-\infty}^{x} \int_{-\infty}^{y} \sum_{i=1}^{\infty} (F(u))^{i-1} \, dF(u) dF(y) = \int_{-\infty}^{x} \int_{-\infty}^{y} \frac{dF(u)}{1-F(u)} dF(y) \\ &= \int_{-\infty}^{x} R(y) \, dF(y) \end{aligned} \tag{1.2}$$

If F(x) has a density f(x), then the probability density function (pdf) of $X_{U(2)}$ is

$$f_2(x) = R(x)\, f(x) \tag{1.3}$$

The distribution function $F_3(x) = P(X_{U(3)} \le x)$

$$= \int_{-\infty}^{x} \int_{-\infty}^{y} \sum_{i=0}^{\infty} (F(u))^i\, R(u)\, dF(u)\, dF(y) = \int_{-\infty}^{x} \int_{-\infty}^{y} \frac{R(u)}{1-F(u)}\, dF(u)\, dF(y)$$

$$= \int_{-\infty}^{x} \frac{(R(u))^2}{2!}\, dF(u) \tag{1.4}$$

The pdf $f_3(x)$ of $X_{U(3)}$ is

$$f_3(x) = \frac{(R(x))^2}{2!} f(x), \quad -\infty < x < \infty \tag{1.5}$$

It can similarly be shown that the distribution function $F_n(x)$ of $X_{U(n)}$ is

$$F_n(x) = P(X_{U(n)} \le x)$$
$$= \int_{-\infty}^{x} f(u_n) du_n \int_{-\infty}^{u_n} \frac{f(u_{n-1})}{1-F(u_{n-1})} du_{n-1} \ldots \int_{-\infty}^{u_2} \frac{f(u_1)}{1-F(u_1)} du_1$$
$$= \int_{-\infty}^{x} \frac{R^{n-1}(u)}{(n-1)!} dF(u), \quad -\infty < x < \infty . \tag{1.6}$$

This can be expressed as

$$F_n(x) = \int_{-\infty}^{R(x)} \frac{u^{n-1}}{(n-1)!} e^{-u}\, du, \quad -\infty < x < \infty$$

$$\bar{F}_n(x) = 1 - F_n(x) = \bar{F}(x) \sum_{j=0}^{n-1} \frac{(R(x))^j}{j!} = e^{-R(x)} \sum_{j=0}^{n-1} \frac{(R(x))^j}{j!}$$

The pdf $f_n(x)$ of $X_{U(n)}$ is

$$f_n(x) = \frac{R^{n-1}(x)}{(n-1)!} f(x), \qquad -\infty < x < \infty . \tag{1.7}$$

The joint pdf $f(x_1, x_2, \ldots, x_n)$ of the n record values $X_{U(1)}, X_{U(2)}, \ldots, X_{U(n)}$) is given by

$$f(x_1, x_2, \ldots x_n) = r(x_1)\, r(x_2) \ldots r(x_{n-1})\, f(x_n) \tag{1.8}$$

$$\text{for } -\infty < x_1 < x_2 < \ldots < x_{n-1} < x_n < \infty,$$

where $r(x) = \frac{d}{dx} R(x) = \frac{f(x)}{1-F(x)}, \quad 0 < F(x) < 1.$

The function r(x) is known as hazard rate.

The joint pdf of $X_{U(i)}$ and $X_{U(j)}$ is

$$f(x_i,x_j) = \frac{(R(x_i))^{i-1}}{(i-1)!} r(x_i) \frac{(R(x_j)-R(x_i))^{j-i-1}}{(j-i-1)!} f(x_j) \tag{1.9}$$

for $-\infty < x_i < x_j < \infty$.

In particular for i = 1 and j = n we have

$$f(x_1,x_n) = r(x_1) \frac{(R(x_n)-R(x_1))^{n-2}}{(n-2)!} f(x_n), \text{ for } -\infty < x_1 < x_n < \infty.$$

The conditional pdf of $X_{U(j)} | X_{U(i)} = x_i$ is

$$f(x_j \mid X_{U(i)} = x_i) = \frac{f_{ij}(x_i,x_j)}{f_i(x_i)} = \frac{(R(x_j)-R(x_i))^{j-i-1}}{(j-i-1)!} \frac{f(x_j)}{1-F(x_i)} \tag{1.10}$$

for $-\infty < x_i < x_j < \infty$.

The marginal pdf of the nth lower record value can be derived by using the same procedure as that of the nth upper record value. If we use H(u) = -ln F(u), 0< F(u) < 1 and $h(u) = -\frac{d}{du}H(u)$, then

$$P(X_{L(n)} \le x) = \int_{-\infty}^{x} \frac{\{H(u)\}^{n-1}}{(n-1)!} dF(u) \tag{1.11}$$

and the corresponding the pdf $f_{(n)}$ can be written as

$$f_{(n)}(x) = \frac{(H(x))^{n-1}}{(n-1)!} f(x). \tag{1.12}$$

The joint pdf of $X_{L(1)}, X_{L(2)}, \ldots, X_{L(m)}$ can be written as

$$f_{(1)(2)\ldots(m)}(x_1, x_2, \ldots, x_m) = h(x_1)h(x_2)\ldots h(x_{m-1}) f(x_m), \quad -\infty < x_m < x_{m-1} < \ldots < x_1 < \infty$$

$$= 0, \text{ otherwise.} \tag{1.13}$$

The joint pdf of $X_{L(r)}$ and $X_{L(s)}$ is

$$f_{(r),(s)}(x,y) = \frac{(H(x))^{r-1}}{(r-1)!} \frac{[H(y)-H(x)]^{s-r-1}}{(s-r-1)!} h(x) f(y) \tag{1.14}$$

for s > r and $-\infty < y < x < \infty$.

Let us consider the standard Rayleigh distribution with pdf

$$f(x) = x\, e^{-x^2/2}, \quad x > 0$$

and cdf

$$F(x) = 1 - e^{-x^2/2}, \quad x > 0. \tag{1.15}$$

Let $\mu_n = E(X_{U(n)})$, $V_{n,n} = \text{var}(X_{U(n)})$ and $V_{m,n} = \text{cov}(X_{U(m)}\, X_{U(n)})$, then

$$\mu_n = \sqrt{2}\,\frac{\Gamma(n+\frac{1}{2})}{\Gamma(n)}, \quad V_{n,n} = 2\left[n - \left(\frac{\Gamma(n+1/2)}{\Gamma(n)}\right)^2\right] \text{ and}$$

$$V_{m,n} = 2\left[\frac{\Gamma(m+1/2)}{\Gamma(m)}\right]\left[\frac{\Gamma(n+1)}{\Gamma(n+1/2)} - \frac{\Gamma(n+1/2)}{\Gamma(n)}\right], \text{ for } 1 \le m < n.$$

Proof:

$$\mu_n = \frac{1}{\Gamma(n)}\int_0^\infty x\{-\ln(1-F(x))\}^{n-1} f(x)\,dx$$

$$= \frac{1}{\Gamma(n)}\int_0^\infty x\left(\frac{x^2}{2}\right)^{n-1} e^{-x^2/2}\, x\,dx = \frac{1}{\Gamma(n)}\sqrt{2}\int_0^\infty u^{1/2}\, u^{n-1}\, e^{-u}\,du$$

$$= \sqrt{2}\,\frac{\Gamma(n+1/2)}{\Gamma(n)}.$$

Similarly it can be shown that

$$\mu_n^2 = E(X_{U(n)}^2) = 2\frac{\Gamma(n+1)}{\Gamma(n)} = 2n$$

$$\mu_{m,n} = \frac{1}{\Gamma(m)\,\Gamma(n-m)}\int_0^\infty\int_0^y xy\left(\frac{x^2}{2}\right)^{m-1} x\left(\frac{y^2}{2}-\frac{x^2}{2}\right)^{n-m-1} y\, e^{-y^2/2}\,dx\,dy$$

$$= \frac{1}{\Gamma(m)\,\Gamma(n-m)}\,\frac{1}{2^{m-1}}\int_0^\infty y\left(\frac{y^2}{2}\right)^{n-m-1} y\, e^{-y^2/2} I_y\,dy,$$

where

$$I_y = \int_0^y (x^2)^m \left(1-\frac{x^2}{y^2}\right)^{n-m-1} dx = \frac{1}{2}\, y^{2m+1}\, B(m+1/2,\, n-m),$$

with $B(a,b) = \dfrac{\Gamma(a)\,\Gamma(b)}{\Gamma(a+b)}$.

On simplification we get

$$V_{n,n} = 2\left[n - \left(\frac{\Gamma(n+1/2)}{\Gamma(n)}\right)^2\right], \text{ and}$$

$$V_{m,n} = 2\left[\frac{\Gamma(m+1/2)}{\Gamma(m)}\right]\left[\frac{\Gamma(n+1)}{\Gamma(n+1/2)} - \frac{\Gamma(n+1/2)}{\Gamma(n)}\right], \text{ for } 1 \le m < n.$$

$$= \left[\frac{\Gamma(m+1/2}{\Gamma(m)}\right]\left[\frac{\Gamma(n)}{\Gamma(n+1/2}\right] V_{n,n} = a_m\, b_n, \text{ say, for } m < n.$$

Then

$$a_m = 2\frac{\Gamma(m+\frac{1}{2})}{\Gamma(m)} \text{ and } b_n = \frac{\Gamma(n+1)}{\Gamma(n+1/2)} - \frac{\Gamma(n+1/2)}{\Gamma(n)}$$

Consider the Gumbel distribution function

$$F(x) = \exp(-e^{-x}), \text{ for all } -\infty < x < \infty. \tag{1.16}$$

Gumbel sistribution is also known as type I extreme value distribution. This distribution has been used in the analysis of data concerning floods, extreme sea levels and air pollution problems . It can be shown that

$$X_{L(m)} \stackrel{d}{=} X - (W_1 + \frac{W_2}{2} + \ldots\ldots + \frac{W_{m-1}}{m-1}), \tag{1.17}$$

where $W_1, W_2, \ldots, W_m$ are independently distributed as negative exponential random variables with mean unity and X is distributed as Gumbel distribution with F(x) as given in (1.16). It can be shown that

For $r < m$

$E(X_{L(r)}) = \upsilon_r^*$, $Var(X_{L(r)}) = V_{r,r}^*$, $r = 1,2, \ldots, m$ and

$Cov(X_{L(r)}, X_{L(m)}) = Var(X_{L(m)})$, $1 \leq r < m$,

with

$$\upsilon_1^* = \upsilon, \; v_{j-1}^* - (j-1)^{-1}, \; j \geq 2, \text{ and}$$

$$V_{1,1}^* = \frac{\pi^2}{6} \text{ and } V_{j,j}^* = V_{j-1,j-1}^* - (j-1)^{-2}, \quad j \geq 2.$$

Here υ is the Euler's constant.

Thus we can write $Cov(X_{L(r)}, X_{L(m)}) = a_r b_m$, where $a_r = 1$ and $b_m = V_{m,m}^*$.

It can similarly be shown that for many univariate distributions including exponential, Pareto, uniform and Weibull, the covariance of any two (mth and nth) upper or lower record values can be expressed as the product of a_m and b_n, where a_m is a function of the parameters of the distribution and m only and b_n is a function of the parametes of the distributions and n only. We will call such a family as a family belonging to class C. We will consider here the estimations of location and scale paramters of the family of distributions belonging to class C.

Let X be a random variable with location and scale parameters as μ and σ ,We will assume that

$$\frac{X-\mu}{\sigma} \in C \qquad (2.1)$$

and X has finite variance. Here we will consider the minimum variance linear unbiased estimators (MVLUE) of μ and σ using upper record values. The MVLUE of μ and σ using lower record values are similar.

(a) Minimum Variance Linear Unbiased Estimators (MVLUE)

Suppose $E\left(\frac{X_{U(n)}-\mu}{\sigma}\right)=\alpha_n$ and $Cov(X_{U(m)}\ X_{U(n)})=a_m\ b_n\sigma^2,\ 1\le m\le n.$

Theorem 2.1.

Let $\hat{\mu}$ and $\hat{\sigma}$ be the minimum variance linear unbiased estimators of μ and σ respectively based on the observed record values $x_{U(1)}, x_{U(2)}, \ldots, x_{U(m)}$. then

$$\hat{\mu}=w_{11}X_{U(1)}+w_{12}X_{U(2)}+\cdots+w_{1m}X_{U(m)}$$

$$\hat{\sigma}=w_{21}X_{U(1)}+w_{22}X_{U(2)}+\cdots+w_{2m}X_{U(m)}$$

with $Var(\hat{\mu})=\frac{e_{12}^{\bullet}}{D_0}\sigma^2$, $Var(\hat{\sigma})=\frac{c_{11}^{\bullet}}{D_0}\sigma^2$ and $Cov(\hat{\mu},\hat{\sigma})=\frac{c_{12}^{\bullet}}{D_0}$

where

$$w_{1k}=\frac{c_{22}^{\bullet}e_{1k}-c_{12}^{\bullet}e_{2k}}{D_0},\ k=1,2,\ldots,m,$$

$$w_{2k}=\frac{c_{11}^{\bullet}e_{2k}-c_{21}^{\bullet}e_{1k}}{D_0},\ k=1,2,\ldots,m,$$

$$D_0=c_{11}^{\bullet}c_{22}^{\bullet}-(c_{12}^{\bullet})^2$$

$$c_{11}^{\bullet}=\sum_{i=1}^{m}c_{i1}^2\quad c_{12}^{\bullet}=c_{21}^{\bullet}=\sum_{i=1}^{m}c_{i1}c_{i2},\ c_{22}^{\bullet}=\sum_{i=1}^{m}c_{i2}^2$$

$$c_{11}=a_{11}=\frac{1}{\sqrt{a_1b_1}},\ c_{12}=\alpha_1a_{11}=\frac{\alpha_1}{\sqrt{a_1b_1}}\ c_{ij}=0,i>2$$

$$c_{k1}=\frac{b_{k-1}-b_k}{\sqrt{b_{k-1}b_k}}\frac{1}{\sqrt{a_kb_{k-1}-a_{k-1}b_k}},\ c_{k2}=\frac{b_{k-1}\alpha_k-b_k\alpha_{k-1}}{\sqrt{b_kb_{k-1}(a_kb_{k-1}-a_{k-1}b_k)}},\ c_{ki}=0,i>2,k=2,\ldots,m.$$

$$C'A=E=\begin{pmatrix} e_{11} & e_{12} & . & e_{1m} \\ e_{21} & e_{22} & . & e_{2m} \end{pmatrix}$$

and

$$e_{11}=\sum_{i=1}^{m}c_{i1}a_{i1}=c_{11}a_{11}+c_{21}a_{21},\ e_{ik}=\sum_{j=1}^{m}c_{ki}a_{jk}=c_{ki}a_{kk}+c_{k+1i}a_{k+1k},\ k=2,3,\ldots,m-1$$

$e_{11} = \sum_{i=1}^{m} c_{i1}\, a_{i1} = c_{11}\, a_{11} \quad + c_{21}\, a_{21}\,, e_{1k} = \sum_{j=1}^{m} c_{k1}\, a_{jk} = c_{k1} a_{kk} + c_{k+11}\, a_{k+1k}$, k = 2,3,..., m-1

$e_{2k} = \sum_{j=1}^{m} c_{k2}\, a_{jk} = c_{k2} a_{kk} + c_{k+12}\, a_{k+1k}$, k = 2,3,..., m-1, $e_{1m} = c_{m1}\, a_{mm}$ and $e_{2m} = c_{m2}\, a_{mm}$.

Proof:

$$V_1 = \frac{1}{\sqrt{a_1 b_1}} X_{U(1)}, V_2 = \sqrt{\frac{b_1}{b_2}}\ \{a_2 b_1 - a_1 b_2\}^{-1} \left[X_{U(2)} - \frac{b_2}{b_1} X_{U(1)} \right]$$

$$V_3 = \sqrt{\frac{b_2}{b_3} \{a_3 b_2 - a_2 b_3\}^{-1}} \left[X_{U(3)} - \frac{b_3}{b_2} X_{U(2)} \right], \ldots\ldots\ldots\ldots,$$

$$V_m = \sqrt{\frac{b_{m-1}}{b_m} \{a_m b_{m-1} - a_{m-1} b_m\}^{-1}} \left[X_{U(m)} - \frac{b_m}{b_{m-1}} X_{U(m-1)} \right]$$

Then we can write

$V = AX$, $X' = (X_{U(1)}, X_{U(2)}, \ldots, X_{U(m)})$,

where

$$A = \begin{pmatrix} a_{11} & a_{12} & \cdots & a_{1m} \\ a_{21} & a_{22} & \cdots & a_{2m} \\ \cdot\cdot & \cdots & \cdots & \cdots \\ a_{m1} & a_{m2} & \cdots & a_{mm} \end{pmatrix} \text{ with } a_{11} = \frac{1}{\sqrt{a_1 b_1}}, a_{1i} = 0, i > 1,$$

$$a_{21} = -\sqrt{\frac{b_2}{b_1}} \frac{1}{\sqrt{a_2 b_1 - a_1 b_2}}, a_{22} = \sqrt{\frac{b_1}{b_2}} \frac{1}{\sqrt{a_2 b_1 - a_1 b_2}}, a_{2i} = 0, i > 2$$

$$a_{k\,k-1} = -\sqrt{\frac{b_k}{b_{k-1}}} \frac{1}{\sqrt{a_k b_{k-1} - a_{k-1} b_k}},\ a_{kk} = \sqrt{\frac{b_{k-1}}{b_k}} \frac{1}{\sqrt{a_k b_{k-1} - a_{k-1} b_k}},\ a_{k,k+i} = 0. i > 0$$

$a_{k,k-i} = 0, i > 1, k = 2,3,\ldots,m.$

In terms of matrix notation, we have

$E(X) = \mu\, 1 + \sigma\, \alpha = B\theta$

where

$\alpha' = (\alpha_1, \alpha_2, \cdots, \alpha_m), \alpha_i = E\left(\frac{X_{U(i)} - \mu}{\sigma} \right), i = 1,2,\ldots,m,\ 1' = (1,1,\ldots,1)$

$$B=\begin{pmatrix} 1 & \alpha_1 \\ 1 & \alpha_2 \\ \cdot & \cdot \\ 1 & \alpha_m \end{pmatrix},\ \theta'=(\mu,\ \sigma).\text{Thus }\ E(V)\ =AB\theta = C\theta.\text{ say.}$$

$$\text{Now } C=\begin{pmatrix} c_{11} & c_{12} \\ c_{21} & c_{22} \\ \cdot & \cdot \\ c_{m1} & c_{m2} \end{pmatrix},\text{ with}$$

$$c_{11}=a_{11}=\frac{1}{\sqrt{a_1 b_1}}, c_{12}=\alpha_1 a_{11}=\frac{\alpha_1}{\sqrt{a_1 b_1}}, c_{1i}=0, i>2$$

$$c_{k1}=\frac{b_{k-1}-b_k}{\sqrt{b_{k-1}b_k}}\frac{1}{\sqrt{a_k b_{k-1}-a_{k-1}b_k}}, c_{k2}=\frac{b_{k-1}\alpha_k-b_k\alpha_{k-1}}{\sqrt{b_k b_{k-1}(a_k b_{k-1}-a_{k-1}b_k)}}, c_{ki}=0, i>2, k=2,\ldots,m.$$

$Var(V)=\sigma^2 I$, where I is an identity matrix of order $m\times m$.

$Var(V_j)=\sigma^2$, $j=1,2,\ldots,m$ and $Cov(V_i, V_j)=0$, $i\neq j$.

Using the ordinary least squares, we get the minimum variance unbiased estimator of θ as

$$\hat{\theta}=(C'C)^{-1}C'V=(C'C)^{-1}C'AX=WX \tag{2.2}$$

Now

$$C'C=\begin{pmatrix} \sum_{i=1}^{m} c_{i1}^2 & \sum c_{i1}c_{i2} \\ \sum c_{i1}c_{i2} & \sum_{i=1}^{m} c_{i2}^2 \end{pmatrix}=\begin{pmatrix} c_{11}^{\bullet} & c_{12}^{\bullet} \\ c_{21}^{\bullet} & c_{22}^{\bullet} \end{pmatrix},\text{ say.}$$

Let D_0 = the determinant of C'C. We have

$D_0=c_{11}^{\bullet}c_{22}^{\bullet}-(c_{12}^{\bullet})^2$, since $c_{12}^{\bullet}=c_{21}^{\bullet}$.

$$\text{Let } C'A=E=\begin{pmatrix} e_{11} & e_{12} & \cdot & e_{1m} \\ e_{21} & e_{22} & \cdot & e_{2m} \end{pmatrix}$$

then we have

$$e_{11}=\sum_{i=1}^{m} c_{i1}a_{i1}=c_{11}a_{11}+c_{21}a_{21},\quad e_{21}=\sum_{i=1}^{m} c_{i2}a_{i1}=c_{12}a_{11}+c_{22}a_{21}$$

$$e_{1k} = \sum_{j=1}^{m} c_{k1}\, a_{jk} = c_{k1}a_{kk} + c_{k+11}\, a_{k+1k}$$, k = 2,3,..., m-1, $e_{1m} = c_{m1}\, a_{mm}$

$$e_{2k} = \sum_{j=1}^{m} c_{k2}\, a_{jk} = c_{k2}a_{kk} + c_{k+12}\, a_{k+1k}$$, k = 2,3,..., m-1 and $e_{2m} = c_{m2}\, a_{mm}$.

Let W = (C'C)$^{-1}$E, then

$$\frac{1}{D_0}\begin{pmatrix} c_{22}^{\bullet} & -c_{12}^{\bullet} \\ -c_{21}^{\bullet} & c_{11}^{\bullet} \end{pmatrix}\begin{pmatrix} e_{11} & e_{12} & \cdots & e_{1m} \\ e_{21} & e_{22} & \cdots & e_{2m} \end{pmatrix}$$

Writing $W = \begin{pmatrix} w_{11} & w_{12} & \cdots & w_{1m} \\ w_{21} & w_{22} & w_{23} & w_{2m} \end{pmatrix}$, we have

$$\hat{\mu} = w_{11}\, X_{U(1)} + w_{12}\, X_{U(2)} + \cdots + w_{1m} X_{U(m)}$$
$$\hat{\sigma} = w_{21}\, X_{U(1)} + w_{22}\, X_{U(2)} + \cdots + w_{2m} X_{U(m)}$$

where

$$w_{1k} = \frac{c_{22}^{\bullet}\, e_{1k} - c_{12}^{\bullet} e_{2k}}{D_0}$$, k = 1,2,...,m

$$w_{2k} = \frac{c_{11}^{\bullet}\, e_{2k} - c_{21}^{\bullet} e_{1k}}{D_0}$$, k = 1, 2,..,m.

with $ar(\hat{\mu}) = \frac{e_{12}^{\bullet}}{D_0}\sigma^2$, $ar(\hat{\sigma}) = \frac{c_{11}^{\bullet}}{D_0}\sigma^2$ and $Cov(\hat{\mu},\hat{\sigma}) = \frac{c_{12}^{\bullet}}{D_0}$.

Rremark 2.1. Suppose $E(X_{L(n)}) = \mu + \alpha_n \sigma$ and $Cov(X_{L(m)} X_{L(n)}) = a_m b_n$, $1 \le m \le n$, then

$$\hat{\mu} = w_{11}\, X_{L(1)} + w_{12}\, X_{L(2)} + \cdots + w_{1m} X_{L(m)}$$
$$\hat{\sigma} = w_{21}\, X_{L(1)} + w_{22}\, X_{L(2)} + \cdots + w_{2m} X_{L(m)}$$

(b) Best Linear Invariant Estimators

Theorem 2

The best linear invariant (in the sense of minimum mean squared error and invariance with respect to the location parameter μ) estimators (BLIE) $\tilde{\mu}$ $\tilde{\sigma}$ of μ and σ are

$$\tilde{\mu} = \hat{\mu} - \hat{\sigma}\left(\frac{E_{12}}{1+E_{22}}\right) \text{ and } \tilde{\sigma} = \hat{\sigma}/(1+E_{22})$$

where $\hat{\mu}$ and $\hat{\sigma}$ are MVLUE of μ and σ and

$$\begin{pmatrix} \mathrm{Var}(\hat{\mu}) & \mathrm{Cov}(\hat{\mu},\hat{\sigma}) \\ \mathrm{Cov}(\hat{\mu},\hat{\sigma}) & \mathrm{Var}(\hat{\sigma}) \end{pmatrix} = \sigma^2 \begin{pmatrix} E_{11} & E_{12} \\ E_{12} & E_{22} \end{pmatrix}$$

The mean squared errors of these estimators are

$\text{MSE}(\tilde{\mu}) = \sigma^2 (E_{11} - E_{11}^2 (1+E_{22})^{-1})$ and $\text{MSE}(\tilde{\sigma}) = \sigma^2 E_{22} (1+E_{22})^{-1}$.

3. Applications

We now give some applications of the Theorem 2.1 to give minimum variance unbiased estimates of some univariate distributions.

3(a) Example 1. Exponential distribtion

Let the pdf of the random variable X be given by $f(x) = \frac{1}{\sigma}\exp\left(-\frac{x-\mu}{\sigma}\right), x \ge 0.$

$$E(X_{U(k)}) = \mu + k\sigma, Var(X_{U(k)}) = k\sigma^2 \text{ and } Cov(X_{U(m)} X_{U(n)}) = m\sigma^2, m < n.$$

Thus $\alpha_k = k, k = 1,2,\ldots, a_m = m$ and $b_n = 1, 1 \le m \le n,$

$$c_{11} = 1, c_{i1} = 0. i > 1,\ c_{i2} = 1, i = 1,2,\ldots,m$$

$$D_0 = m-1, c_{11}^{\bullet} = m, c_{12}^{\bullet} = c_{21}^{\bullet} = -1 \text{ and } c_{22}^{\bullet} = \ ,$$

$$e_{11} = 1, e_{1i} = 0, i > 1, e_{2i} == 0, i = 1,2,\ldots, \text{m}-1, e_{2m} = \frac{\text{m}}{\text{m}-1},$$

$$w_{11} = \frac{m}{m-1}, w_{1i} = 0, i = 2,\cdots,m-1, w_{1m} = -\frac{1}{m}$$

$$w_{21} = -\frac{1}{m-1}, w_{2i} = 0, i = 2,\cdots,m-1, w_{2m} = \frac{1}{m}$$

Thus we have the MVLUE estimators $\hat{\mu}$ and $\hat{\sigma}$ of μ and σ respectively based on the observed record values $x_{U(1)}, x_{U(2)}, \ldots, x_{U(m)}$ as

$$\hat{\mu} = \left(\text{m}X_{U(1)} - X_{U(m)}\right)/(\text{m}-1)$$
$$\hat{\sigma} = \left(X_{U(m)} - X_{U(1)}\right)/(\text{m}-1)$$

with

$$\text{Var}(\hat{\mu}) = \text{m}\sigma^2/(\text{m}-1), \text{Var}(\hat{\sigma}) = \sigma^2/(\text{m}-1) \text{ and } \text{Cov}(\hat{\mu},\hat{\sigma}) = -\sigma^2/(\text{m}-1).$$

Using $E_{11} = \frac{m}{m-1}, E_{12} = -\frac{1}{m-1}$ and $E_{22} = \frac{1}{m-1}$, we obtain the best linear invariant (in the sense of minimum mean squared error and invariance with respect to the location parameter μ) estimators (BLIE) $\tilde{\mu}$ $\tilde{\sigma}$ of μ and σ as

$\tilde{\mu} = \left\{(\text{m}+1)X_{U(1)} - X_{U(m)}\right\}/\text{m}$ and $\hat{\sigma} = \left(X_{U(m)} - X_{U(1)}\right)/m$ with

$Var(\tilde{\mu}) = \frac{m+1}{m}\sigma^2$ and $\text{Var}(\hat{\sigma}) = \frac{\text{m}-1}{\text{m}^2}\sigma^2$

$$Var(\tilde{\mu}) = \frac{m+1}{m}\sigma^2 \text{ and } Var(\hat{\sigma}) = \frac{m-1}{m^2}\sigma^2$$

By direct calculation these results were obtained by Ahsanullah (1980).

3(b) Example 2. Uniform distribution.

Suppose X is distributed uniformly over the interval (μ, μ +σ), then

$$E(X_{U(k)}) = \mu + (1-2^{-k})\sigma, Var(X_{U(k)}) = (3^{-k} - 4^{-k})\sigma^2, k = 1,\ldots,m \text{ and}$$
$$Cov(X_{U(k)}, X_{U(m)}) = 2^{k-m} Var(X_{U(k)}), 1 \le k < m.$$

Thus

$$\alpha_k = 1 - \frac{1}{2^k}, a_k = 2^k\left[3^{-k} - 4^{-k}\right] \text{and } b_k = 2^{-k}, k = 1,\ldots,m.$$

$$a_1 b_1 = \frac{1}{3} - \frac{1}{4} = \frac{1}{12}, \quad a_{11} = \frac{1}{\sqrt{a_1 b_1}} = 2\sqrt{3}, \ a_{i1} = 0, i >$$

$$a_k b_{k-1} - a_{k-1} b_k = 2^k (3^{-k} - 4^{-k}) \cdot \frac{1}{2^{k-1}} - 2^{k-1}(3^{-(k-1)} - 4^{-(k-1)}) \cdot \frac{1}{2^k} = \frac{1}{2(3^k)}$$

$$a_{kk-1} = -\sqrt{\frac{b_{k-1}}{b_k}} \frac{1}{\sqrt{a_k b_{k-1} - a_{k-1} b_k}} = -\frac{\sqrt{2(3^k)}}{\sqrt{2}} = -3^{\frac{k}{2}}$$

$$a_{kk} = \sqrt{\frac{b_k}{b_k - 1}} \frac{1}{\sqrt{a_k b_{k-1} - a_{k-1} b_k}} = \sqrt{2}\,(\sqrt{2(3^{\frac{k}{2}})} = 2(3^{\frac{k}{2}})$$

$$c_{11} = a_{11} = 2\sqrt{3}, \ c_{12} = \alpha_1 a_{11} = \sqrt{3}, c_{1i} = 0, i > 2$$

$$e_{21} = \sum_{i=1}^{m} c_{i2} a_{i1} = c_{12} a_{11} + c_{22} a_{21} = \sqrt{3}(2\sqrt{3}) + 3(3) = -3$$

$$c_{k1} = \frac{b_{k-1} - b_k}{\sqrt{b_{k-1} b_k}} \frac{1}{\sqrt{a_k b_{k-1} - a_{k-1} b_k}} = 3^{\frac{k}{2}}$$

$$c_{k2} = \frac{b_{k-1}\alpha_k - b_k \alpha_{k-1}}{\sqrt{b_{k-1} b_k}} \frac{1}{\sqrt{a_k b_{k-1} - a_{k-1} b_k}} = 3^{\frac{k}{2}}, k = 2,3,\ldots,m.$$

$$e_{11} = \sum_{i=1}^{m} c_{i1} a_{i1} = c_{11} a_{11} + c_{21} a_{21} = (2\sqrt{3})(2\sqrt{3}) - 3(3) = 3$$

$$e_{1k} = \sum_{j=1}^{m} c_{j1} a_{jk} = c_{k1} a_{kk} + c_{k+11} a_{k+1k} = -3^{k+1}, \ k = 2,3,\ldots, m-1$$

$$e_{2k} = \sum_{j=1}^{m} c_{k2} a_{jk} = c_{k2} a_{kk} + c_{k+12} a_{k+1k}, = -3^{k+1}, k = 2,3,\ldots, m-1$$

$$e_{1m} = c_{m1} a_{mm} = 2(3^m) \text{ and } e_{2m} = c_{m2} a_{mm} = 2(3^m).$$

Let

$$W=(C'C)^{-1}E = \frac{1}{D_0}\begin{pmatrix} c_{22}^{\bullet} & -c_{12}^{\bullet} \\ c_{21}^{\bullet} & c_{11}^{\bullet} \end{pmatrix}\begin{pmatrix} e_{11} & e_{12} & \cdots & e_{1m} \\ e_{21} & e_{22} & \cdots & e_{2m} \end{pmatrix},$$

where $D_0 = 3T$, $c_{11}^{\bullet} = 12+T$, $c_{12}^{\bullet} = 6+T$, $c_{22}^{\bullet} = 3+T$ and

$$T = 3^2 + 3^3 + \cdots + 3^m = \frac{9(3^{m-1}-1)}{2}.$$

Writing $W = \begin{pmatrix} w_{11} & w_{12} & \cdots & w_{1m} \\ w_{21} & w_{22} & w_{23} & w_{2m} \end{pmatrix}$, we have on simplification

$$w_{11} = \frac{9+2T}{T}, w_{1k} = \frac{3^k}{T}, k = 2,3,\cdots,m-1, w_{1m} = -\frac{2(3^m)}{T}$$

$$w_{21} = -\frac{18+T}{T}, w_{2k} = -2w_{2k}, k = 2,3,\cdots,m.$$

On simplification, we obtain

$$\hat{\mu} = X_{U(1)} - \frac{1}{2}\hat{\sigma}$$

$$\hat{\sigma} = -\frac{18+2T}{T}X_{U(1)} - \frac{2(3^2)}{T}X_{U(2)} - \cdots - \frac{2(3^{m-1})}{T}X_{U(m-1)} + \frac{4(3^m)}{T}X_{U(m)}$$

$$\mathrm{Var}(\hat{\mu}) = \frac{3+T}{3T}\sigma^2, \mathrm{Cov}(\hat{\mu},\hat{\sigma}) = -\frac{6+T}{3T}\sigma^2 \textit{ and } \mathrm{Var}(\hat{\sigma}) = \frac{12+T}{3T}\sigma^2.$$

The BLIE estimators $\tilde{\mu}$ $\tilde{\sigma}$ of μ and σ are respectively

$$\tilde{\mu} = \hat{\mu} + \frac{6+T}{12+4T}\hat{\sigma} \text{ and } \tilde{\sigma} = \frac{3T}{12+4T}\hat{\sigma} \text{ with}$$

$$MSE(\tilde{\mu}) = \sigma^2\frac{4+T}{12+4T} \text{ and } MSE(\tilde{\sigma}) = \sigma^2\frac{12+T}{12+4T}.$$

Theorems 1 and 2 can be used for the MVLUE and BLIE for all the members of the class C . The class C includes Weibull distribution, Extreme value distributions, Pareto distributions,etc.

4. Distributions with conditional expectations of adjacent record values as linear .

For many distributions including exponential, Pareto and uniform

$$E(X_{U(n)} \mid X_{U(n-1)} = y) = a + by \qquad (4.1)$$

for some constants a and b . We will say a rv X with distribution function F belongs to the class C_1 if its n th record value satisfy the condition (4.1).. In this paper, we will consider the record values of random variables belonging to class C_1.

4.1. MAIN RESULTS

RESULT 4.1.

If the sequence of rvs $X_1, X_2, \ldots$, belong to class C_1 with finite variance, then

$$\mathrm{Cov}(X_{U(n)}, X_{U(m)}) = b^{n-m}\,\mathrm{Var}(X_{U(m)}),\ n > m.$$

Proof:

$$\begin{aligned} E(X_{U(m+2)}) &= EE(X_{U(m+2)} \mid X_{U(m+1)} = t) \\ &= EE(a + bt \mid X_{U(m)} = y) \\ &= E(a + b(a+by)) \\ &= a+ab + b^2 E(X_{U(m)}). \end{aligned}$$

In general

$$E(X_{U(n)}) = a + ab + ab^2 + a b^3 + \ldots + ab^{n-m-1} + b^{n-m} E(X_{U(m)}).$$

$$= a\frac{b^{n-m}-1}{b-1} + b^{n-m} E(X_{U(m)}),\ if\ b \neq 1$$

$$= (n-m)\,a + E(X_{U(m)}),\ \text{if } b = 1.$$

Thus

$$\mathrm{Cov}(X_{U(m)}, X_{U(n)}) = b^{n-m}\,\mathrm{Var}(X_{U(m)}),\ \text{if } b \neq 1$$

$$= \mathrm{Var}(X_{U(m)}),\ \text{if } b = 1.$$

RESULT 4.2.

If the sequence of i.i.d. rvs $X_1, X_2, \ldots$ has an absolutely continuous distribution function F with support on [c, d), where c is finite and d may be infinite and has finite expectation. Further we assume that F belongs to the class C_1 with b > 0 and with d $= \infty\ if\ b \geq 1\ and\ d = -\frac{a}{b-1}\ if\ b < 0$

Then $1 - F(x) = \left(\frac{a+(b-1)c}{a+(b-1)x}\right)^{\frac{b}{b-1}}\ for\ b \neq 1$

and $1 - F(x) = e^{-x/a}$ for b = 1,

if and only if $E(X_{U(n)} | X_{U(n-1)} = y) = a + b y$.

Proof:

Writing the conditional expectation of $X_{U(n)} | X_{U(n-1)} = y)$ and simplifying we get

$$a + b\,y = y + \int_y^d \frac{1-F(x)}{1-F(y)}\,dx \qquad (4.2.1)$$

Differentiating both sides of the above equation with respect to y, we obtain

$$b = (a + (b-1)\,y)\,(f(y)/(1-F(y))) \qquad (4.2.2)$$

Finally integrating (2.2) with respect to y from c to x, we obtain the result.

It can be shown that for the rv X having the distribution function as given in the Result 4.22,

The expected value and the variance of X are respectively $E(X) = a + bc$ and $Var(X) = (b(a + (b-1)c)^2)/(2-b)$.

For various results of the rv X based record values for the case b =1, see Ahsanullah ((1988),(1980)). The results for the case b >1 are similar to those for the case b < 1. In this paper, we will consider the results corresponding to b >1. The distribution function corresponding to b > 1 was introduced by Pickands (1975) in connection to extreme value distribution.

For inference based on record values for the generalized extreme value distribution Ahsanullah and Holland (1994).

Let f_n be the probability density function of the n th record value, $X_{U(n)}$. Then

$$f_n(x) = \frac{\{\frac{b}{b-1}\ln\frac{a+(b-1)c}{a+(b-1)x}\}^n}{n!}\,\frac{b(a+(b-1)c)^{\frac{b}{b-1}}}{(a+(b-1)x)^{\frac{2b-1}{b-1}}} \qquad (4.2.3)$$

It can be shown from (4.2.3) that

$$E(X_{U(n)}) = (1/(b-1))(b^n(a + (b-1)c) - a)$$

$$Var(X_{U(n)}) = (b-1)^{-2}(a + (b-1)c)^2\, b^n((2-b)^{-n} - b^n).$$

and $\quad Cov(X_{U(m)}, X_{U(n)}) = b^{n-m}\, Var(X_{U(m)})$.

We will assume without the loss generality the lower bound, c of the rv X as zero and $(Y - \mu)/\sigma = X$, then

$$E(Y) = \mu + a\,\sigma \text{ and } Var(Y) = a^2 b(2 - b)^{-1} \sigma^2.$$

For the variance to be finite, b must be less than 2.

Let $T_1, T_2, \ldots, T_n$ be the record values of Y corresponding to $X_{U(1)}, X_{U(2)}, \ldots, X_{U(n)}$.

It can be shown that

$$T_n = \mu - \frac{a\sigma}{b-1} + \frac{a\sigma}{b-1}\prod_{i=1}^{n} U_i$$

where $U_1, U_2, \ldots, U_n$ are independent and identically distributed with

$$P(U_i \le x) = 1 - x^{-b/(b-1)}$$

Thus

$$E(T_n) = m + a\frac{b^n - 1}{b-1}\sigma, \; Var(T_n) = a^2(b-1)^{-2} b^n \{(2-b)^{-n} - b^n\}\sigma^2$$

and $\quad Cov(T_m, T_n) = b^{n-m}\, Var(T_m)$, $m < n$.

We can write the Variances and Covariances of T's as

$$Var(T_r) = a_r b_r \sigma_1^2 \text{ and } Cov(T_r, T_s) = a_r b_s \sigma_1^2, \; r \le s,$$

where

$$a_r = [(2-b)^{-r} - b^r],\; b_r = b^r,\; r = 1, 2, \ldots, \text{ and } \sigma_1^2 = a^2 (b-1)^{-2} \sigma^2.$$

There are other distributions (see Balakrisnan, Ahsanullah and Chan(1992)) for which $Cov(T_r, T_s)$ can be factored out as the product of two factors, one depends on r and the parameters and the other depends on s and the parameters.

ESTIMATORS OF μ AND σ.

The minimum Variance linear unbiased estimator (MVLUE) of $\hat{\mu}$, $\hat{\sigma}$ of μ and σ are

$$\hat{\mu} = T_1 - a\hat{\sigma}$$

and

$$\hat{\sigma} = a^{-1}[\{\frac{b-1}{b} - D^{-1}(\frac{2-b}{b})^3\}T_1 + D^{-1}\frac{b-1}{b}\sum_{i=2}^{n-1}(\frac{2-b}{b})^{i+1}T_i + D^{-1}(\frac{2-b}{b})^{n+1}T_n$$

where

$$D = \sum_{i=2}^{n}(\frac{2-b}{b})^{i+1}$$

Proof.

Let $T' = (T_1, T_2, \ldots, T_n)$, then we can write

$$E(T) = m L + d\,\sigma_1$$

$L' = (1,1,\ldots,1)$, $d' = (d_1, d_2, \ldots, d_n)$ and $d_i = b^n - 1$, $i = 1,2,\ldots, n$.

Let $V(T) = \sigma_1^2\,\Sigma$, $\Sigma^{-1} = W$, $W = (V^{i,j})$

It can be shown that $V^{i+1,i} = V^{i,i+1} = -\dfrac{b}{(1-b)^2}(\dfrac{2-b}{b})^{i+1}$

$$V^{i,i} = \frac{1+2b-b^2}{(1-b)^2}(\frac{2-b}{b})^i, \quad i = 1,2,\ldots,n-1$$

$$V^{n,n} = \frac{1}{(1-b)^2}(\frac{2-b}{b})^n$$

$V^{i,j} = 0$, if $|i-j| > 1$

Let $W_i = a^{-1}((2-b)/b)^{i/2}(T_i - b\,T_{i-1})$, $i = 1,2,\ldots,n$ and $T_0 = 0$. Then $\mathrm{Var}(W_i) = \sigma^2$ and $\mathrm{Cov}(W_i, W_k) = 0$, $i \neq k$, $1 \le i, k \le n$.

Suppose $W' = (W_1, W_2, \ldots, W_n)$ and $E(W) = Aq$, where $q' = (\mu, \sigma)$

$A' = [A_1\ A_2]$, $A_1' = (d_1, d_2, \ldots, d_n)$, $A_2' = (e_1, e_2, \ldots, e_n)$,

$d_i = ((2-b)/b)^i(1-b)$, $e_i = d_i/(1-b)$, $i = 2,3,\ldots,n$, $d_1 = (1/a)((2-b)/b)^{1/2}$ and $e_1 = a\,d_1$.

Using least squares estimation method, we get on simplification

$$\hat{\mu} = T_1 - a\,\hat{\sigma}$$

$$\hat{\sigma} = a^{-1}\,[\{\frac{b-1}{b} - D^{-1}(\frac{2-b}{b})^3\}T_1 + D^{-1}\frac{b-1}{b}\sum_{i=2}^{n-1}(\frac{2-b}{b})^{i+1}T_i + D^{-1}(\frac{2-b}{b})^{n+1}T_n$$

$$\text{Var}(\hat{\mu}) = (\frac{a^2 T}{b^2 D})\sigma^2$$

$$\text{Var}(\hat{\sigma}) = \{(\frac{2-b}{b})^2 + (\frac{b-1}{b})^2 T\}\,\sigma^2 / D$$

$$\text{cov}(\hat{\mu},\hat{\sigma}) = \{(\frac{b-1}{b})T - \frac{2-b}{b}\}\frac{a\sigma^2}{bD}$$

where $T = \sum_{i=1}^{n}(\frac{2-b}{b})^i$ and $D = \sum_{i=2}^{n}(\frac{2-b}{b})^{i+1}$

Let a =2 and b = 1.5, then

$$\hat{\mu} = (17/12)\,T_1 - (1/4)\,T_2 - (1/12)\,T_3 - (1/12)\,T_4$$

and

$$\hat{\sigma} = -(5/8)\,T_1 + (1/8)T_2 + (1/24)T_3 + (1/24)T_4$$

The corresponding variance and covariance are

$$\text{Var}(\hat{\mu}) = \frac{160}{9}\sigma^2,\ \text{Var}(\hat{\sigma}) = \frac{40}{81}\sigma^2 \text{ and } \text{Cov}(\hat{\mu},\hat{\sigma}) = -\frac{41}{9}\sigma^2$$

BEST LINEAR INVARIANCE ESTIMATORS (BLIE)

The best linear invariant (in the sense of minimum mean squared error and invariance with respect to the location parameter m) estimators

$\tilde{\mu}, \tilde{\sigma}$ *of* μ *and* σ are

$$\tilde{\mu} = \hat{\mu} - \hat{\sigma}\left(\frac{E_{12}}{1+E_{22}}\right)$$

$$\text{and } \tilde{\sigma} = \hat{\sigma}\,(1+E_{22})^{-1}$$

where $\hat{\mu}$ *and* $\hat{\sigma}$ *are MVLUE of* μ *and* σ

and

$$\begin{pmatrix} \text{var}(\hat{\mu}) & \text{cov}(\hat{\mu},\hat{\sigma}) \\ \text{cov}(\hat{\mu},\hat{\sigma}) & \text{var}(\hat{\sigma}) \end{pmatrix} = \sigma^2 \begin{pmatrix} E_{11} & E_{12} \\ E_{12} & E_{22} \end{pmatrix}$$

The mean squared errors of these estimators are

$$MSE\ (\tilde{\mu}) = \sigma\ (E_{11} - E_{12}{}^2\ (1 + E_{22})^{-1}\ )$$

$$MSE\ (\tilde{\sigma}) = \sigma^2\ E_{12}\ (1 + E_{22})^{-1}\ .$$

Substituting the values of E_{11}, E_{12}, E_{22}, we get

$$\tilde{\mu} = \hat{\mu} - \frac{b\{(b-1)T - 2 + b\}}{T}\hat{\sigma}$$

and

$$\hat{\sigma} = \hat{\sigma}\frac{D\,b^2}{T}$$

$$MSE\ (\tilde{\mu}) = \frac{a^2\ \sigma^2}{b^2\ D}[T - \frac{\{(b-1)T - (2-b)\}^2}{T}]$$

$$MSE\ (\tilde{\sigma}) = \frac{a\sigma^2}{T}\{(b-1)T - (2-b)]$$

With n = 4 and b = 1.5, we have

$$\tilde{\mu} = \frac{2105}{1920}T_1 - \frac{37}{640}T_2 - \frac{37}{1920}T_3 - \frac{37}{1920}T_4$$

$$\tilde{\sigma} = -\frac{1125}{1200}T_1 + \frac{9}{160}T_2 + \frac{3}{160}T_3 + \frac{3}{160}T_4$$

$$MSE(\tilde{\mu}) = \frac{127413}{10800}\sigma^2 \text{ and } MSE(\tilde{\sigma}) = \frac{121}{160}\sigma^2.$$

PREDICTOR of T_s

We shall consider the prediction of T_s based on n observed record values for s > n.

Let $H' = (h_1, h_2, \ldots, h_n)$, where $\sigma^2 h_i = Cov(T_i, T_s)$, i = 1,2,...,n and

$\mu_s = \sigma^{-1} E(T_s - \mu)$. The best linear unbiased predictor (BLUP) of T_s is

$\hat{T}_s$ *, where*

$$\hat{T}_s = \hat{\mu} + \hat{\sigma}\gamma_s + H'V^{-1}(T - \hat{\mu}L - \hat{\sigma}\gamma)$$

Now

$H'V^{-1} = (0,0,\ldots, b^{s-n})$ and

$$\hat{T}_s = \hat{\mu} + \hat{\sigma}\gamma_s + b^{s-n}(T_n - \hat{\mu} - \hat{\sigma}\gamma_n$$

$$= b^{s-n} T_n + (1 - b^{s-n})\hat{\mu} + (1 - b^{s-n})\gamma_s \hat{\sigma}$$

The best(unrestricted) least squares predictor of T_s is $T_s^{*} = E(T_s | T_1, T_2, \ldots, T_n)$.

Thus $T_s^{*} = = \mu + a\frac{b^{s-n} - 1}{b - 1}\sigma + ab^{s-n}(T_n - \mu)$

If we substitute the MVLUE of m and s , then T_s^{*} becomes $\hat{T}_s$.

Let $\tilde{T}_s$ be the best linear invariant predictor of T_s.

From the results of Mann (1969) it follows that

$$\tilde{T}_s = \hat{T}_s - \frac{c^{*}_{12}}{1 + E_{22}}\hat{\sigma}$$

where

$$c^{*}_{12} = \text{Cov}(\hat{\sigma}, (1 - H'V^{-1}L)\hat{\mu} + (\gamma_s - H'V^{-1}\delta)a\,\hat{\sigma})$$

and $\quad -H'V^{-1}L = 1 - b^{s-n}$ *and* $\gamma_s - H'V^{-1}\delta = \gamma_{s-n}$

Thus

$$\tilde{T}_s = \hat{T}_s - \gamma_{s-n}\frac{2 - b}{b^2}$$

Considering the MSE of the predictor, it can be shown that

$$MSE(T_s^{*}) \le MSE(\tilde{T}_s) \le MSE(\hat{T}_s)$$

REFERENCES

Ahsanullah, M. (1978). Record values and the exponential distribution. Ann. Inst. Statist. Math. 30,429-433.

Ahsanullah, M. (1979). Characterization of the exponential distribution by record values. Sankhya, 41, B, 116-121.

Ahsanullah, M. (1980). Linear prediction of record values for the two parameter exponential distribution. Ann. Inst. Stat. Math. 32, A, 363-368.

Ahsanullah, M. (1981). Record values of the exponentially distributed random variables. Statistiche Hefte, 2, 121-127.

Ahsanullah, M. (1982). Characterizations of the exponential distribution by some properties of record values. Statistiche Hefte, 23, 326-332.

Ahsanullah, M. (1986 b). Estimation of the parameters of a rectangular distribution by record values. Comp. Stat. Quarterly 2, 119-125.

Ahsanullah, M. (1988). Introduction to Record Statistics, Ginn Press, Needham Heights, MA.

Ahsanullah, M. (1990). Estimation of the parameters of the Gumbel distrbution based on m record values. Comp. Statist. Quarterly, 3,231- 239.

Ahsanullah, M. (1992). Record values of independent and identically distributed continuous random variables. Pak. J. Statist.8(2),A, 9-34.

Ahsanullah,M. and Holland,B.(1994).On the use or record values to estimate the location and scale parameters of the generalized extreme value distribution. Sankhya 56, Series A, pp. 480-499.

Ahsanullah,M. (1994) Record values, random record models and concommitants. J.Stat.Res. 28, 89-109.

Ahsanullah, M. (1995). Record Statistics. Nova Science Publishers, Inc. Commack, NY, USA.

N. Balakrishnan and Ahsanullah,M.(1995). Recurrence Relations for single and product moments of record values from exponential distribution (1995). J. Appl. Statist. Scs.2(1), 73-88.

N. Balakrishnan and Ahsanullah,M.(1995). Relations for Single and product moments of record values from Lomax distribution (1995). Sankhya 56 B(2) 140-146.

N. Balakrishnan and Ahsanullah,M. and C. Chan (1995). Recurrrence relation of the moments of the Logistic distribution.(1995). Journal of Applied Statistical Science, 2(3) ,233-248.

Arnold, B. C., Balakrishnan, N., and Nagaraja, H. N. (1992). A first course in Order Statistics. Wiley, New York.

Balakrishnan, N., Ahsanullah, M. and Chan, P.S. (1992). Relations for single and product moments of record values from Gumbel distribution.Stat.and Prob. Letters 15,227 223-227.

Balakrishnan, N. and Ahsanullah, M. and Chan,P.S.(1992) Recurrence relations for moments of record values from generalized extreme value distribution. Comm. Statist. Theory and Methods. 22(5),1471-1482.

Balakrishnan, N. and Ahsanullah, M. (1994) Recurrrence relation of the moments of the Generalized Pareto distribution. Comm. in Stat.- Theor. & Meth. 23 (10), 2841-2852. Joint with Dr. N. Balakrishnan

Chandler, K. N. (1952). The distribution and frequency of record values. J. R. Statist. Soc. B 14, 220-228.

Feller, W. (1966). An Introduction to Probability Theory and its Applications. Vol. II, Wiley, New York.

Goldberger, A. S. (1962). Best Linear Unbiased Predictors in the Generalized Linear Regression Model. J. Amer. Statist. Assoc. 57, 369-375.

Mann, N.R. (1969). Optimum estimators for linear functions of location and scale parameters. Ann. Math. Statist. 40, 2149-2155.

Nagaraja, H. N. (1988) Record values and related statistics -- a review, Commun. Statist. Theory Meth. 17, 2223-2238.

Nevzorov, V. B. (1988). Records. Theo. Prob. Apl. 32, 201-228.

Applied Statistical Science, II
ISBN 1-56072-469-2

A TRANSFER THEOREM FOR MULTITYPE PROCESSES AND APPLICATIONS

Ibrahim Rahimov*

Abstract

The multitype, continuous time and indecomposable Markov branching process with immigration is considered. Using the martingale approach a limit theorem proved for such processes, when the totality of immigrating individuals at a given time depends on evo lution of the processes generating by immigrated before individuals, under the assumption that the limit theorem for corresponding process without immigration holds. Corollaries of the limit theorem obtained for the cases of finite and infinite second mom ents of offspring distribution in critical processes and examples of applications of these results are provided.

Key words: MULTITYPE, INDECOMPOSABLE, CRITICALITY, DEPENDENT IMMIGRATION, STOPPING TIME.
AMS 1991 SUBJECT CLASSIFICATION: PRIMARY 60J80, SECONDARY 60G70.

1 Introduction

Let for any $t \in [0, \infty)$ the piar $(\theta_k^t, \mathbf{Y}_k^t)$ be some n-variate point process, where θ_k^t are random variables with values from the set $R_+ = [0, \infty)$ and $\mathbf{Y}_k^t =$

* Addresses: School of Mathematical Sciences University of Science Malaysia, 11800, Penang, Malaysia and Institute of Mathematics, Hodjaev St.,29, 700143, Tashkent, Uzbekistan.

$(Y_{k1}^t, ..., Y_{kn}^t)$ be random vectors taking "values" from $N^n, N = \{1, 2, ...\}, t \in R_+$. Here θ_k^t and $\mathbf{Y}_k^t$ are the time and "size" of kth jump of the process

$$\mathbf{Y}^t(u) = (Y_1^t(u), ..., Y_n^t(u)) = \sum_{k \geq 1} \chi(\theta_k^t \leq u)\mathbf{Y}_k^t$$

respectively and $\chi(A)$ is the indicator function of an event A.

Suppose that the $\mathbf{Y}_k^t$ is a random vector of immigrating at time θ_k^t individuals of types $T_1, T_2, ..., T_n$. These immigrating individuals generate independent and identically distributed n-type continuous time Markov br anching stochastic processes.

If we enumerate simultaneously immigrating individuals of the type T_j by $i = 1, 2, ...,$ then the triple (k, i, j) corresponds to the ith individual of the type T_j, $j = 1, 2, ..., n$, immigrating at time θ_k^t. We shall call the multitype branching process generated by individual (k, i, j) as "(k, i, j)-process".

We denote

$$\mathbf{X} = \{\mathbf{X}_{ki}^j(t) = (X_{ki}^{j1}(t), X_{ki}^{j2}(t), ..., X_{ki}^{jn}(t)), k \geq 1, i \geq 1, j = 1, 2, ..., n\} \quad (1)$$

the family of all possible (k, i, j)-processes. Here $X_{ki}^{jm}(t)$ is the number of individuals of type T_m in the (k, i, j)-process at time t. Then the branching process with n-types and immigration is defined by $\mathbf{Z}(t) = (Z_1(t), Z_2(t), ..., Z_n(t))$, $t \geq 0, \mathbf{Z}(0) = \mathbf{O}$, where

$$Z_m(t) = \sum_{j=1}^{n} \sum_{k=1}^{N(t)} \sum_{i=1}^{Y_{kj}^t} X_{ki}^{jm}(t - \theta_k^t) \quad (2)$$

is the number of individuals of type T_m at time t and $N(t)$ is the number of jumps of the point process $\mathbf{Y}^t(u)$ up to time t, that is

$$N(t) = \sum_{k:\theta_k^t \leq t} 1.$$

Multitype branching processes with immigration have been studied widely in the literature (see Quine(1970), Kaplan(1974), Shurenkov(1976)). One can find a sufficiently full bibliography of such papers in the books Mode(1971), Sevast'yanov(1971), Jagers(1975), Badalbaev and Rahimov(1993) and in the review of Vatutin and Zubkov(1993). However, the independence of processes of reproduction and immigration was assumed in these publications. In the described above model this assumption means that the family $\mathbf{X}$ of independent and identically distributed branching processes and point processes $(\theta_k^t, \mathbf{Y}_k^t)$ are independent. Under this assumption the study of $\mathbf{Z}(t)$ can be reduced to the analyses of a relation for its generating function. If we do not assume the independence, it is not possible to get an explicit expression for the generating function of the process.

On the other hand in real branching processes often the immigration process depends on reproduction. For instance, if we consider the process of urban population growth, the number of immigrants at present depends on the lives of past immigrants and their descendants. Another example is the neutron chain reaction in a nuclear reactor with an external neutron source. If one wants to support such a process by immigration, it is apparent that the immigration process depends on reproduction.

Here the process $\mathbf{Z}(t)$ will be considered without the assumption of independence of processes of reproduction and immigration. We prove a limit theorem for $\mathbf{Z}(t)$ under the assumption that the limit theorem for corresponding process without immigration holds. Such theorems are called "transfer theorems" for branchin g processes (see Rahimov (1995)). Corollaries of the transfer theorem will be obtained for cases of finite and infinite second moments of offspring distribution in critical processes and examples of applications of these results will be discussed. Single- type branching processes with reproduction-dependent immigration were studied in Rahimov(1992, 1995).

Let $\Re_{k,i}^j(t)$ be the σ-algebra generated by the evolution of the (k,i,j)-process up to time t. We have not spoken on the independence of (k,i,j)-processes and the immigration process until now. Presuppose now that the following basic assumptions are fulfilled:
(a) the family $\{\theta_k^t, k \in N, t \in R_+\}$ is measurable with respect to some σ-algebra $\Re_0$, which is independent of the family $\Re_{ki}^j(t)$.
(b) for any $r = 0, 1, ...$ and $p = 1, 2, ..., n$,

$$\{Y_p(k,t) \leq r\} \in \Im_{kr}(t), \tag{3}$$

where

$$\Im_{kr}(t) = \prod_{l=1}^{k-1} \prod_{q=1}^{n} \prod_{i=1}^{Y_q(l,t)} \Re_{li}^q(t) \times \prod_{q=1}^{n} \prod_{i=1}^{r} \Re_{ki}^q(t) \times \Re_0.$$

The direct products of the random number of σ-algebras we shall understand as

$$\prod_{i=1}^{Y} \Re_i = \{A : A \bigcap \{Y = j\} \in \prod_{i=1}^{j} \Re_i\}, \quad \prod_{i=1}^{0} \Re_i = \{\emptyset, \Omega\}.$$

Under the condition (3) the totality of immigrating individuals at time θ_k^t may depend on evolution of the processes generated by the individuals which immigrated up to time θ_k^t. Note also that for processes with reproduction-independent immigration the condition (3) is fulfille d with $\mathbf{Y}_k^t$ are $\Re_0$-measurable and in this case σ-algebras $\Re_{ki}^j(t)$ and $\Re_0$ are independent.

2 The transfer theorem and corollaries

For n-dimesional vectors $\mathbf{x} = (x_1, x_2, ..., x_n), \mathbf{y} = (y_1, y_2, ..., y_n)$ we denote $\mathbf{x} \oplus \mathbf{y} = (x_1y_1, ..., x_ny_n), \mathbf{x}^{\mathbf{y}} = (x_1^{y_1}, ..., x_n^{y_n}), (\mathbf{x}, \mathbf{y}) = x_1y_1 + ... + x_ny_n, \mathbf{1} = (1, 1, ..., 1), \mathbf{O} = (0, 0, ..., 0), \mathbf{e} = (e, e, ..., e)$ and $\mathbf{x} \geq \mathbf{y}$ or $\mathbf{x} > \mathbf{y}$ if $x_i \geq y_i$ or $x_i > y_i, i = 1, 2, ..., n$, respectively.

We also denote for $j = 1, 2, ..., n$,

$$F_j(t, \mathbf{S}) = ES_1^{X_{ki}^{j1}(t)} S_2^{X_{ki}^{j2}(t)} \dots S_n^{X_{ki}^{jn}(t)}, \quad \mathbf{S} = (S_1, S_2, ..., S_n)$$

the generating functions of the (k, i, j)-processes.

Assume that

$$\sup_t EX_{ki}^{jl}(t) \leq C_0 < \infty,$$

and for any fixed $\lambda = (\lambda_1, ..., \lambda_n) > \mathbf{O}$ and for some non-increasing functions $\mathbf{Q}(t) = (Q_1(t), ..., Q_n(t))$

$$\lim_{t \to \infty} \frac{1 - F_j(t, \mathbf{e}^{-\lambda \oplus \mathbf{Q}(t)})}{Q_j(t)} = 1 - \varphi(\lambda), \tag{4}$$

where $\varphi(\lambda) = \varphi(\lambda_1, ..., \lambda_n)$ is the Laplace transform of a random vector having finite expectation, $\mathbf{Q}(t) \to \mathbf{O}$ and for any $x \in [0, 1]$,

$$\lim_{t \to \infty} \frac{Q_j(t)}{Q_j(tx)} = \pi(x) \tag{5}$$

and this convergence is uniform in each interval of the form $[\varepsilon, 1)$ for any $\varepsilon > 0$ and $\pi(x)$ is a continuous function for $x \in (0, 1]$.

Conditions (4) and (5) can be satisfied for critical or close to critical (in the case of transition phenomena) multitype branching processes. The limit function $\pi(x)$ in (5) necessarily has the form x^{α} for some $\alpha \in [0,\infty)$ (see S.I Resnick(1987), p.14).

For the immigration process we assume that

$$\sum_{j=1}^{n} Q_j(t) Y_j^t(tx) \xrightarrow{P} T(x), \tag{6}$$

as $t \to \infty$ for $0 \le x \le 1$, where $T(x)$ is some $\Re_0$- measurable, stochastically continuous for $x = 1$ stochastic process with non-decreasing trajectories, $T(0) = 0$ and $T(1) < \infty$ almost everywhere.

Theorem. *If conditions (3) - (6) are satisfied, then*

$$\mathbf{W}(t) = \mathbf{Q}(t) \oplus \mathbf{Z}(t) \xrightarrow{\mathcal{D}} \mathbf{W} = (W_1, W_2, ..., W_n),$$

where

$$Ee^{-(\lambda,\mathbf{W})} = E\exp\left\{-\int_0^1 \frac{1-\varphi(\lambda_1\pi(1-x), ..., \lambda_n\pi(1-x))}{\pi(1-x)} dT(x)\right\}$$

(the value of the integrand at $x = 1$ is defined by continuity).

We denote P_α^j, $j = 1, 2, ..., n$, the infinitesimal probabilities of the (k,i,j)-process, that is

$$P\{\mathbf{X}_{ki}^j(t) = \alpha\} = \delta_\alpha^{e_j} + P_\alpha^j t + o(t), \quad t \downarrow 0$$

where $\alpha = (\alpha_1, \alpha_2, ..., \alpha_n)$, $\alpha_i \in \mathcal{N}_0 = \{0, 1, ...\}, e_j = (\delta_j^1, ..., \delta_j^n), \delta_j^i$ is the Kronecer's delta and δ_β^α for any vectors α and β is defined as following:

$$\delta_\beta^\alpha = \begin{cases} 1 & \text{for } \alpha = \beta \\ 0 & \text{for } \alpha \ne \beta. \end{cases}$$

We also denote

$$F^j(\mathbf{S}) = \sum_{\alpha \in \mathcal{N}_0^n} P_\alpha^j S_1^{\alpha_1} ... S_n^{\alpha_n}, \quad \mathbf{F}(\mathbf{S}) = (F^1(\mathbf{S}), ..., F^n(\mathbf{S})), F^j(\mathbf{1}) = 0.$$

Let for $i, j, k = 1, 2, \cdots, n$

$$a_i^j = \frac{\partial F^j(\mathbf{S})}{\partial S_i} |_{\mathbf{S}=\mathbf{1}}, b_{ik}^j = \frac{\partial^2 F^j(\mathbf{S})}{\partial S_i \partial S_k} |_{\mathbf{S}=\mathbf{1}}$$

be finite, $\mathbf{A} = \left\| a_i^j \right\|$ be the matrix of derivatives, ρ be its Peron root and the right and the left eigenvectors $\mathbf{U} = (u_1, u_2, ..., u_n)$ and $\mathbf{V} = (v_1, v_2, ..., v_n)$ corresponding to the Peron root be such that

$$\mathbf{AU} = \rho \mathbf{U}, \quad \mathbf{VA} = \rho \mathbf{V}, \quad \sum_{i=1}^{n} u_i v_i = 1, \ \sum_{i=1}^{n} u_i = 1.$$

If $\mathbf{A}$ is indecomposable, aperiodic and $\rho = 0$, then the limit theorem for the critical multitype Markov branching processes holds (see Sevast'yanov(1971), p.214, for example). In this case the conditions (4) and (5) are satisfied with

$$\begin{aligned}
\varphi(\lambda) &= (1+\hat{\lambda})^{-1}, \hat{\lambda} = \sum_{j=1}^{n} \lambda_j u_j v_j, \\
Q_j(t) &= P\{\mathbf{X}_{ki}^j(t) \neq \mathbf{0}\} \sim \frac{2u_j}{bt}, \qquad t \to \infty, \\
\pi(x) &= x,
\end{aligned}$$

where $b = \sum_{j,m,k=1}^{n} v_j b_{mk}^j u_m u_k$ and we have the following result.

Corollary 1. *If conditions (3) and (6) are satisfied, then*

$$\lim_{t\to\infty} P\left\{\frac{2u_j Z_j(t)}{bt} \le y_j, j = 1, 2, ..., n\right\} = A(\mathbf{y}),$$

where the distribution $A(\mathbf{y})$, $\mathbf{y} = (y_1, y_2, \ldots, y_n)$, *has the Laplace transform*

$$\int_{R_+^n} e^{-(\mathbf{y},\lambda)} dA(\mathbf{y}) = E\exp\left\{-\hat{\lambda}\int_0^1 (1+(1-x)\hat{\lambda})^{-1} dT(x)\right\}.$$

Let now the generating functions $F^j(\mathbf{S})$ such that

$$\sum_{j=1}^{n} v_j F^j(\mathbf{1} - \mathbf{U}x) = x^{1+\alpha} L(x), \tag{7}$$

where $0 < x \leq 1, \alpha \in (0,1]$, and $L(x)$ is a slowly varying function as $x \downarrow 0$. In this case the following limit theorem for the critical multitype branching process holds (see Vatutin (1977)).

Proposition. *Under the condition (7) we have a)*

$$P\{\mathbf{X}^j_{ki}(t) \neq \mathbf{0}\} \sim u^j t^{-1/\alpha} L_1(t)$$

as $t \to \infty$, where $L_1(t)$ is a slowly varying as $t \to \infty$ function; b)

$$\lim_{t\to\infty} P\left\{X^{jl}_{ki}(t) q(t) \leq x_l v_l, l = 1,2,...,n \mid \mathbf{X}^j_{ki}(t) \neq \mathbf{0}\right\} = G(\mathbf{x}),$$

where

$$q(t) = \sum_{j=1}^{n} v_j (1 - F^j(t, \mathbf{0}) \sim t^{-1/\alpha} L_1(t),$$

and $G(\mathbf{x}) = G(x_1, x_2, ..., x_n)$ a distribution having the Laplace transform

$$\int_{R^n_+} e^{-(\mathbf{x},\lambda)} dG(\mathbf{x}) = 1 - (1 + \bar{\lambda}^{-\alpha})^{-1/\alpha}, \bar{\lambda} = \sum_{j=1}^{n} \lambda_j.$$

It follws from the Proposition 1 that under the assumption (7) conditions (4) and (5) are satisfied with

$$\begin{aligned} \varphi(\lambda) &= 1 - (1 + (\lambda)^{-\alpha}\hat{)}^{-1/\alpha}, \\ Q_j(t) &= P\{\mathbf{X}^j_{ki}(t) \neq \mathbf{0}\} \sim \frac{u_j L_1(t)}{t^{1/\alpha}}, \qquad t \to \infty, \\ \pi(x) &= x^{1/\alpha}, \end{aligned}$$

and in this case we have the folowing corollary.

Corollary 2. *If conditions (3), (6) and (7) are satisfied, then*

$$\lim_{t\to\infty} P\left\{Q_j(t) Z_j(t) \leq y_j, j = 1,2,...,n\right\} = B(y_1, y_2, ..., y_n),$$

where the distribution $B(\mathbf{y}), \mathbf{y} = (y_1, y_2, ..., y_n)$ has the Laplace transform

$$\int_{R^n_+} e^{-(\mathbf{y},\lambda)} dB(\mathbf{y}) = E \exp\left\{- \int_0^1 [1 - x + \hat{\lambda})^{-\alpha}]^{-1/\alpha} dT(x)\right\}$$

Corollaries 1 and 2 give examples of branching processes for which conditions (4) and (5) of the theorem are fulfilled. Now we consider some examples of immigration processes satisfying condition (6).

Example 1. Let $\mathbf{Y}_k^t \equiv \mathbf{Y}_k$ and for $r = 0, 1, ..., p = 1, 2, ..., n$

$$\{Y_{kp} \leq r\} \in \mathcal{B}_r(k,t) = \prod_{i=1}^{r} \prod_{q=1}^{n} \Re_{ki}^{q}(t).$$

Since (k,i,j)–processes are independent, the vector $\mathbf{Y}_k$ and processes $\{\mathbf{X}_{li}^j(t),$ $l, i \geq 1,\ l \neq k, j = 1, ..., p\}$ are also independent. If $\mathbf{Y}_k, k = 1, 2, ...$ have the same distribution, and $N(t)$ is a Pisson process wit h the intensity ν, then $\mathbf{Y}(t) = (Y_1(t), ..., Y_n(t))$ is the multivariate compound Poisson process. Therefore the Laplace transform of $Y_j(t)$ have the following form

$$Ee^{-uY_j(t)} = \exp\{\nu t(f_j(u) - 1)\},$$

where $f_j(u)$ the Laplace transform of Y_{kj}. Using this representation it is not difficult to show that, if $E\mathbf{Y}_k = \mathbf{a} = (a_1, a_2, ..., a_n)$ is finite, then

$$\frac{Y_j(t)}{t} \xrightarrow{P} \nu a_j$$

In fact, since

$$\lim_{t\to\infty} t(1 - f_j(\frac{u}{t})) = ua_j,$$

we have that $E\exp\{-ut^{-1}Y_j(t)\}$ tends to $\exp\{-\nu a_j u\}$ as $t \to \infty$ for any $u > 0$. So $t^{-1}Y_j(t)$ converges to $a_j\nu$ in distribution.

Hence, if $\rho = 0$ and $b \in (0, \infty)$, then the condition (6) holds with $T(x) = 2x\nu b^{-1} \sum_{j=1}^{n} a_j u_j$. It is not difficult to see that in this case the Laplace transform in Corollary 1 is

$$(1 + \hat{\lambda})^{-\alpha}, \quad \alpha = 2\nu b^{-1} \sum_{j=1}^{n} a_j u_j, \hat{\lambda} = \sum_{j=1}^{n} \lambda_j u_j v_j.$$

Therefore we have the following result which is a generalization of the well-known theorem on convergence to the gamma distribution.

Corollary 3. *If* $\rho = 0, b \in (0, \infty)$, *the coordinates of the vector* $\mathbf{Y}_{kp}, p =$ $1, 2, ..., n$ *are stopping times with respect to the family* $\mathcal{B}_r(k,t), r = 0, 1, ...,$

it has the same distribution for different k, $\mathbf{a} < \infty$ *and* $N(t)$ *is a Poisson process with the density* ν*, then*

$$\mathbf{W}(t) = (\frac{2u_j Z_j(t)}{bt}, j = 1, 2, ..., n) \xrightarrow{\mathcal{D}} \mathbf{W} = (u_1 v_1 W_1, ..., u_n v_n W_n)$$

where $W_i = W_j$ *with probability 1 and* W_i *has the gamma distridution of the parameter* α.

Remark. There are many theorems on convergence to the gamma distribution obtained for critical single or multitype branching processes with reproduction-independent immigration in the literature. These results require the same moment assumptions tha t the Corollary 3. Thus, it follows from Corollary 3, that the theorem on convergence to the gamma distribution holds, without additional conditions, when the totality of the number of immigrants is the stopping time with respect to the family of σ generated by evolution of reproduction processes.

Example 2. Let again coordinates of the vector $\mathbf{Y}_k$ are stopping times with respect to $\mathcal{B}_r(k,t)$. If $\rho = 0, b \in (0, \infty)$ and

$$\frac{1}{t}\sum_{k=0}^{t} \mathbf{Y}_k \xrightarrow{P} \eta = (\eta_1, ..., \eta_n),$$

as $t \to \infty$, where η is some random vector, then the condition (6) holds with $T(x) = 2xb^{-1}(\mathbf{U}, \eta)$ and the Laplace transform in Corollary 1 is

$$E[(1 + \hat{\lambda})^{-2T/b}], \qquad T = (\mathbf{U}, \eta).$$

If, in addition, $\mathbf{Y}_k, k = 1, 2, ...$ have the same distribution, then $W_k = (\mathbf{U}, \mathbf{Y}_k), k = 1, 2, ...$ are independent and identically distributed. Therefore, if $N(t)$ is a Poisson process, then the process

$$\sum_{j=1}^{n} v_j^{-1} \mathbf{Y}_j(t) = \sum_{k=1}^{N(t)} W_k$$

is compound Poisson. Then the Laplace transform of T is infintely divisible as a limit of positive infinitely divisible Laplace transforms. Therefore it can be written in the following form (see Feller (1967), Sec. 7, Chap. XII)

$$Ee^{-uT} = \exp\left\{-\int_0^\infty \frac{1 - e^{-ux}}{x} P(dx)\right\},$$

where $P(x)$ is a measure such that $\int_0^\infty x^{-1}P(dx) < \infty$. Since

$$E[(1+\hat{\lambda})^{-2T/b}] = E\exp\{-\frac{2T}{b}\log(1+\hat{\lambda})\},$$

using this fact we obtain the following result.

Corollary 4. *If coordinates of* $\mathbf{Y}(k)$ *are stopping times with respect to the family* $\mathcal{B}_r(k,t), r = 0,1,...,$ *then, under the assumptions mentioned above* $\mathbf{W}(t) \xrightarrow{\mathcal{D}} \mathbf{W}, t \to \infty$, *where*

$$Ee^{-(\lambda,\mathbf{W})} = \exp\left\{-\int_0^\infty \frac{(1+\hat{\lambda})^{2x/b}-1}{x(1+\hat{\lambda})^{2x/b}}P(dx)\right\}.$$

Remark. Note that the condition $b \in (0,\infty)$ is not necessary in corollaries 3 and 4. It is clear from arguments above, that one could obtain similar limit theorems replacing this condition by a regularly variation assumption of the tail of th e offspring distribution. For example, we could use the assumption (7) instead of finiteness of b. Of course the family of limiting distributions in this case would be broader than in corollaries 3 and 4.

3 The proof of the transfer theorem

First we consider the function

$$H(t,\lambda) = \prod_{k=1}^{N(t)} \prod_{j=1}^{n} [F_j(t-\theta_k, e^{-\lambda\oplus\mathbf{Q}(t)})]^{Y_{kj}^t}$$

and we shall prove the following lemma.

Lemma 1. *If conditions (4), (5) and (6) are satisfied, then*

$$H(t,\lambda) \xrightarrow{P} H(\lambda) = \exp\left\{-\int_0^1 \frac{1-\varphi(\pi(1-x)\lambda)}{\pi(1-x)}dT(x)\right\}. \tag{8}$$

Proof. Since $Y_{kj}^t = \Delta Y_j^t(\theta_k^t) = Y_j^t(\theta_k^t) - Y_j^t(\theta_k^t - 0)$ and

$$\log H(t, \lambda) = \sum_{j=1}^{n} \sum_{k=1}^{N(t)} Y_{kj}^t \log F_j(t - \theta_k^t, \mathbf{e}^{-\lambda \oplus \mathbf{Q}(t)}),$$

we get

$$H(t, \lambda) = \exp\left\{\sum_{j=1}^{n} \int_0^t \log F_j(t - u, \mathbf{e}^{-\lambda \oplus \mathbf{Q}(t)}) dY_j^t(u)\right\}.$$

We choose $\varepsilon \in (0, 1)$, put $a = 1 - \varepsilon$ and consider

$$A_1 = \sum_{j=1}^{n} \int_0^{ta} \log F_j(t - u, \mathbf{e}^{-\lambda \oplus \mathbf{Q}(t)}) dY_j^t(u). \tag{9}$$

If we denote

$$\pi_j(t, x) = \frac{Q_j(t)}{Q_j(tx)}, \quad \pi(t, x) = (\pi_1(t, x), ..., \pi_n(t, x)), C = C_u(t) = 1 - \frac{u}{t},$$

then for sufficiently large t and $0 \le u \le ta$

$$0 < \pi(\varepsilon) - \varepsilon_1 \le \pi_j(t, C) \le 1 \tag{10}$$

for some $\varepsilon_1 > 0$ and $j = 1, 2, ..., n$. Therefore it follows from (10) and the condition (4) that

$$\sup_{1 \le k \le [at]} \left| \frac{1 - F_j(t - u, \mathbf{e}^{-\lambda \oplus \mathbf{Q}(t)})}{Q_j(t - u)} - 1 + \varphi((\lambda, \pi(t, C))\right| \to 0 \tag{11}$$

as $t \to \infty$. On the other hand, since $\varepsilon \le_u (t) < 1$ for $0 \le u \le ta$, it follows from (5) that

$$\sup_{0 \le u \le ta} |\pi_j(t, C) - \pi(C)| \to 0 \tag{12}$$

as $t \to \infty$ for $j = 1, 2, ..., n$. We obtain from (11) and (12) the following representation

$$-\frac{\log F_j(t - u, \mathbf{e}^{-\lambda \oplus \mathbf{Q}(t)})}{Q_j(t)} = \frac{1 - \varphi(\pi(\theta)\lambda)}{\pi(\theta)}(1 + \alpha_j(u, t)), \tag{13}$$

where $\alpha_j(u,t) \to 0$ for $j = 1, 2, ..., n$, as $t \to \infty$ uniformly with respect to $o \le u \le ta$.

It is not difficult to see that the integral

$$\int_0^{1-\varepsilon} \frac{1-\varphi(\pi(1-x)\lambda)}{\pi(1-x)} d\xi_t(x), \tag{14}$$

where $\xi_t(x) = \sum_{j=1}^n Q_j(t) Y_j^t(tx)$, conveges in probability to

$$B(\lambda, \varepsilon) = \int_0^{1-\varepsilon} \frac{1-\varphi(\pi(1-x)\lambda)}{\pi(1-x)} dT(x)$$

due to condition (6). Since $T(x)$ is stochastically contiuous at $x = 1$ and $\varphi(\lambda)$ is the Laplace transform of a random vector having finite expectation, one can show by standard arguments that the last integral converges in probability to the $B(\lambda, 0)$ as $\varepsilon \to 0$. Hence it follows from (13) that A_1 converges in probability to $-B(\lambda, 0)$ as $t \to \infty$ and $\varepsilon \to 0$.

We now consider

$$A_2 = \sum_{j=1}^n \int_{ta}^t \log F_j(t-u, \mathbf{e}^{-\lambda \oplus \mathbf{Q}(t)}) dY_j^t(u).$$

Using the simple inequality $\log(1-x) \ge -x - x^2/(1-x), 0 \le x < 1$, we get

$$0 > A_2 \ge -\sum_{j=1}^n \int_{ta}^t (R_j(u) + \frac{R_j^2(u)}{1-R_j(u)}) dY_j^t(u),$$

where

$$R_j(u) = R_j(u, \lambda) = 1 - F_j(t-u, \mathbf{e}^{-\lambda \oplus \mathbf{Q}(t)}).$$

It is not difficult to see that

$$\sum_{j=1}^n \int_{ta}^t R_j(u) dY_j^t(u) \le C_0 \sum_{j=1}^n \lambda_j [\xi_t(1) - \xi_t(1-\varepsilon)]$$

and the last difference in probability converges to $T(1) - T(1-\varepsilon)$ as $t \to \infty$. Using again the stochastic continuity of $T(x)$ at $x = 1$ we obtain that this difference in probability converges to 0 as $\varepsilon \to 0$, and, therefore, A_2 converges in probability to 0 as $t \to \infty$ and $\varepsilon \to 0$. The lemma is proved.

Let $\mathbf{X}_i(n) = (X_{i1}(n), X_{i2}(n), ..., X_{ip}(n)), i = 1, 2, ...$ be non-negative p-dimensional random vectors such that $\mathbf{X}_i(n)$ is $F_i(n)$-measurable, where $F_i(n)$ are some σ-algebras such that $F_i(n) \subseteq F_{i+1}(n), i = 0, 1, 2, ...$ for any n. We consider the following sum

$$S_n = \sum_{i=1}^{n} \nu_i(n)\mathbf{X}_i(n), \tag{15}$$

where $\nu_i(n)$ are random variables taking values 0 and 1 and $F_{i-1}(n)$-measurable. Denote

$$f_j^{(n)}(\lambda) = E[e^{-(\lambda, \mathbf{X}_j(n))}|F_{j-1}(n)], \lambda = (\lambda_1, \lambda_2, ..., \lambda_p) \in R_+^p.$$

The following lemma for $p = 1$ was proved in Rahimov(1995, p.29) using a semimartingale technique. A similar result for simple random sums can be seen in Beska et. al.(1982).

Lemma 2. *Let for any $j = 1, 2, ..., n$ random variable $\nu_j(n)$ is $F_{j-1}(n)$ -measurable and*

$$\prod_{j=1}^{n} \left\{ f_j^{(n)}(\lambda) \right\}^{\nu_j(n)} \xrightarrow{P} \varphi(\lambda), n \to \infty, \tag{16}$$

where $\varphi(\lambda)$ is some $\mathcal{F}_0$-measurable random variable, such that $\varphi(\lambda) > 0$ almost everywhere, $\mathcal{F}_0 \subset F_0(n)$ for all $n = 1, 2, ...$ Then

$$E\left[e^{-(\lambda, \mathbf{S}_n)}|\mathcal{F}_0\right] \xrightarrow{P} \varphi(\lambda), n \to \infty. \tag{17}$$

Remark. It should be noted that Lemma 2 is important in our considerations. It allows us to use the Laplace transform techniques in sums of the form (15) with dependent variables. As it was asserted above one could obtain this result using the semim artingale arguments, as in Rahimov (1995). However to have a direct proof of the lemma seems preferable for the further development of the theory. For instance the study of processes with decreasing or increasing reproduction-dependent immigration needs i n results sort of Lemma 2 and it would be difficult to use the semimartingale technique in such processes.

Proof. Let $\widetilde{\mathbf{X}}_i(n) = \mathbf{X}_i(n)\chi(\mathcal{A}_i(n))$, where

$$\mathcal{A}_i(n) = \left\{ \prod_{j=1}^{i} \left\{ f_j^{(n)}(\lambda) \right\}^{\nu_j(n)} \geq \frac{1}{2}\varphi(\lambda) \right\},$$

and $\chi(.)$ is the indicator function. It is clear that $\mathcal{A}_i(n) \in F_{i-1}(n)$. First we shall prove that for any n and $m \leq n$

$$E\left[\prod_{k=1}^{m} Z_k^{\nu_k(n)}(n)|\mathcal{F}_0\right] = 1, \tag{18}$$

where

$$Z_k(n) = \frac{e^{-(\lambda,\widetilde{\mathbf{X}}_k(n))}}{\widetilde{f}_k^{(n)}(\lambda)}, \quad \widetilde{f}_k^{(n)}(\lambda) = E[e^{-(\lambda,\widetilde{\mathbf{X}}_k(n))}|F_{k-1}(n)].$$

If $m = 1$, then

$$E\left[\prod_{k=1}^{1} Z_k^{\nu_k(n)}(n)|\mathcal{F}_0\right] = E[E[Z_1^{\nu_1(n)}(n)|F_0(n)]|\mathcal{F}_0].$$

Using the simple equality

$$e^{-(\lambda,\tilde{\mathbf{X}}_k(n))\nu_k(n)} = \chi(\nu_k(n) = 1)e^{-(\lambda,\tilde{\mathbf{X}}_k(n))} + \chi(\nu_k(n) = 0) \tag{19}$$

we have

$$E\left[\prod_{k=1}^{1} Z_k^{\nu_k(n)}(n)|\mathcal{F}_0\right] = 1$$

Let now (18) is true for $m = i$. Then, since

$$E\left[\prod_{k=1}^{i+1} Z_k^{\nu_k(n)}(n)|\mathcal{F}_0\right] = E\left[E\left[\prod_{k=1}^{i+1} Z_k^{\nu_k(n)}(n)|F_i(n)\right]|\mathcal{F}_0\right],$$

the last expectation can be written as

$$E\left[\prod_{k=1}^{i} Z_k^{\nu_k(n)}\{f_{i+1}^{(n)}(\lambda)\}^{-\nu_{i+1}(n)}E\left[e^{-(,\tilde{\mathbf{X}}_{i+1}(n))\nu_{i+1}(n)}|F_i(n)\right]|\mathcal{F}_0\right].$$

If we use (19) again, we have

$$E\left[e^{-(\lambda,\tilde{\mathbf{X}}_{i+1}(n))\nu_{i+1}(n)}|F_i(n)\right] = \left\{\tilde{f}_{i+1}^{(n)}(\lambda)\right\}^{\nu_{i+1}(n)}.$$

And, therefore

$$E\left[\prod_{k=1}^{i+1} Z_k^{\nu_k(n)}(n)|\mathcal{F}_0\right] = E\left[\prod_{k=1}^{i} Z_k^{\nu_k(n)}(n)|\mathcal{F}_0\right] = 1.$$

Hence (18) holds for any $m = 1, 2, ...$

Now we shall prove that

$$\prod_{i=1}^{m} \left\{\tilde{f}_i^{(n)}(\lambda)\right\}^{\nu_i(n)} \geq \frac{1}{2}\varphi(\lambda). \tag{20}$$

almost everywhere, for all n and $m \leq n$. Since $\mathcal{A}_i(n) \in F_{i-1}(n)$, it is not difficult to find that

$$\tilde{f}_i^{(n)}(\lambda) = E\left[e^{-(\lambda,\tilde{\mathbf{X}}_i(n))}|F_{i-1(n)}\right] = \left\{f_i^{(n)}(\lambda)\right\}^{\chi(\mathcal{A}_i(n))}.$$

If we use this relation with $i = 1$, we have

$$\left\{\tilde{f}_1^{(n)}(\lambda)\right\}^{\nu_1(n)} = \left\{f_1^{(n)}(\lambda)\right\}^{\nu_1(n)\chi(\mathcal{A}_1(n))}$$

which is equal to

$$\left\{f_1^{(n)}(\lambda)\right\}^{\nu_1(n)} \chi(\mathcal{A}_1(n)) + \chi(\bar{\mathcal{A}}_1(n)) \geq \frac{1}{2}\varphi(\lambda)\chi(\mathcal{A}_1(n)) + \chi(\bar{\mathcal{A}}_1(n)).$$

Since $0 \leq \varphi(\lambda) \leq 1$ (as the limit of a sequence with members are ≤ 1), we obtain

$$\left\{\tilde{f}_1^{(n)}(\lambda)\right\}^{\nu_1(n)} \geq \frac{1}{2}\varphi(\lambda)\chi(\mathcal{A}_1(n)) + \frac{1}{2}\varphi(\lambda)\chi(\bar{\mathcal{A}}_1(n)) = \frac{1}{2}\varphi(\lambda).$$

Therefore the relation (20) holds for $m = 1$. Let now it holds for $m = k$. Then we have:

$$\prod_{i=1}^{k+1} \left\{\tilde{f}_i^{(n)}(\lambda)\right\}^{\nu_i(n)} = \prod_{i=1}^{k} \left\{\tilde{f}_i^{(n)}(\lambda)\right\}^{\nu_i(n)} \left\{\tilde{f}_{k+1}^{(n)}(\lambda)\right\}^{\nu_{k+1}(n)},$$

where

$$\left\{\tilde{f}_{k+1}^{(n)}(\lambda)\right\}^{\nu_{k+1}(n)} = \left\{f_{k+1}^{(n)}(\lambda)\right\}^{\nu_{k+1}(n)} \chi(\mathcal{A}_{k+1}(n)) + \chi(\bar{\mathcal{A}}_{k+1}(n)).$$

Therefore

$$\prod_{i=1}^{k+1}\left\{\tilde{f}_i^{(n)}(\lambda)\right\}^{\nu_i(n)}=\prod_{i=1}^{k+1}\left\{\tilde{f}_i^{(n)}(\lambda)\right\}^{\nu_i(n)}\chi(\mathcal{A}_{k+1}(n))+\prod_{i=1}^{k}\tilde{f}_i^{(n)}(\lambda)\chi(\tilde{\mathcal{A}}_{k+1}(n)).$$

Taking into account the definition of $\mathcal{A}_k(n)$ and the induction assumption, we odtain that the last sum is not less than

$$\frac{1}{2}\varphi(\lambda)\chi(\mathcal{A}_{k+1}(n))+\frac{1}{2}\varphi(\lambda)\chi(\bar{\mathcal{A}}_{k+1}(n)=\frac{1}{2}\varphi(\lambda).$$

Hence (20) is true for any m.

Putting $\tilde{\mathbf{S}}_n=\sum_{k=1}^{n}\tilde{\mathbf{X}}_k(n)\nu_k(n)$ and using (18) and (20) we have

$$|E[e^{-(\lambda,\tilde{\mathbf{S}}_n)}|\mathcal{F}_0]-\varphi(\lambda)|\leq\frac{2}{\varphi(\lambda)}E[W_n(\lambda)|\mathcal{F}_0], \tag{21}$$

where

$$W_n(\lambda)=\left|\prod_{k=1}^{n}\left\{\tilde{f}_k^{(n)}(\lambda)\right\}^{\nu_k(n)}-\varphi(\lambda)\right|.$$

The estimate (21) shows that in order to be

$$E\left[e^{-(\lambda,\tilde{\mathbf{S}}_n)}|\mathcal{F}_0\right]\xrightarrow{P}\varphi(\lambda),n\to\infty, \tag{22}$$

it is sufficient that $W_n(\lambda)$ converges in probability to zero as $n\to\infty$. In fact, then we obtain from the dominated convergence theorem that $EW_n(\lambda)$ tends to zero and, since

$$P\{E[W_n(\lambda)|\mathcal{F}_0]>\varepsilon\}\leq\frac{1}{\varepsilon}EW_n(\lambda),$$

we have that $E[W_n(\lambda)|\mathcal{F}_0]$ converges to zero in probability as $n\to\infty$.

For any $\varepsilon>0$ we have

$$P\{W_n(\lambda)>\varepsilon\}\leq P\{V_n(\lambda)>\varepsilon\}+P\{\bigcup_{k=1}^{n}R_k(n)\bigcap\{\nu_k(n)=1\}\}, \tag{23}$$

where

$$V_n(\lambda)=\left|\prod_{k=1}^{n}f_k^{(n)}(\lambda)^{\nu_k(n)}-\varphi(\lambda)\right|,R_k(n)=\left\{f_k^{(n)}(\lambda)\neq\tilde{f}_k^{(n)}(\lambda)\right\}.$$

Since $R_k(n) = \bar{A}_k(n)$, we have to show that

$$P\{T_n\} \to 0, n \to \infty, \tag{24}$$

where

$$T_n = \bigcup_{k=1}^{n} \bar{A}_k(n) \bigcap \{\nu_k(n) = 1\}.$$

We have from the definition of $\mathcal{A}_k(n)$, that

$$\bar{A}_k(n) \subseteq \left\{ \prod_{j=1}^{n} \left\{ f_j^{(n)}(\lambda) \right\}^{\nu_j(n)} < \frac{1}{2}\varphi(\lambda) \right\}, k \leq n.$$

Therefore we have

$$P\{T_n\} \leq P\left\{ \varphi(\lambda) - \prod_{j=1}^{n} \{f_j^{(n)}(\lambda)\}^{\nu_j(n)} > \frac{1}{2}\varphi(\lambda) \right\}.$$

Choosing ε such that $\varepsilon < \varphi(\lambda)$, we obtain due to condition (16)

$$P\{T_n\} \leq P\{V_n(\lambda) > \frac{\varepsilon}{2}\} \to 0, n \to \infty. \tag{25}$$

Thus (24) holds. It follows from (23) and (24) that $W_n(\lambda)$ converges to zero in probability.

It remains to show that the variable $\tilde{\mathbf{S}}_n$ in (22) can be replaced by S_n. It is not difficult to see that

$$E\left[e^{-(\lambda,\tilde{\mathbf{S}}_n)} | \mathcal{F}_0\right] = E\left[e^{-(\lambda,\mathbf{S}_n)} | \mathcal{F}_0\right] + E\left[\left(e^{-(\lambda,\tilde{\mathbf{S}}_n)} - e^{-(\lambda,\mathbf{S}_n)}\right) \chi(T_n) | \mathcal{F}_0\right]$$

and the variable $\chi(T_n)$ converges to zero in probability due to (25).

The lemma is proved.

Proof of Theorem 1. It follows from (2) that $\mathbf{Z}(t)$ can be written in the form

$$\mathbf{Z}(t) = \sum_{k=0}^{N(t)} \sum_{r=1}^{n} \sum_{i=1}^{\infty} \chi(i \leq Y_{kr}^t) \mathbf{X}_{ki}^r (t - \theta_k^t). \tag{26}$$

Let $\{M_t\}$ and $\{N_t^r\}, t \geq 0$ be a sequences of integers such that

$$P\{N(t) > M_t\} \to 0, \quad P\{\bigvee_{k=1}^{N(t)} Y_{kr}^t > N_t^r\} \to 0 \tag{27}$$

as $t \to \infty$ for $r = 1, 2, ..., n$. Putting $N_t = \bigvee_{r=1}^{n} N_t^r$, we define process $\mathbf{W}^*(t)$ by the following relation

$$\mathbf{W}^*(t) = \sum_{l=1}^{k(t)} \nu_l(t)\mathbf{V}_l(t) \tag{28}$$

where $k(t) = nN_tM_t$,

$$\nu_l(t) = \chi(i \le Y_{kr}^t)\chi(k \le N(t)), \quad \mathbf{V}_l(t) = \mathbf{X}_{ki}^r(t-k)$$

for such l that

$$l = N_t n(k-1) + j, \quad j = N_t(r-1) + i \tag{29}$$

and $1 \le j \le nN_t, 1 \le i \le N_t, 1 \le k \le M_t, 1 \le r \le n$. Then we have from (26) that

$$T_t\mathbf{Z}(t) = T_t\mathbf{W}^*(t), \tag{30}$$

where

$$T_t = \chi\{\bigcap_{r=1}^{n}\{\bigvee_{k=1}^{t} Y_{kr}^t \le N_t\}\}\chi(N(t) \le M_t).$$

It is not difficult to see that for any l satisfying (29) the vector $\mathbf{V}_l(t)$ is $\mathcal{F}_{ki}(t)$-measurable. On the other hand it follows from condition (3) that $\nu_l(t)$ is $\mathcal{F}_{ki-1}(t)$-measurable. Thus Lemma 2 is applicable to $\mathbf{W}^*(t)$. Note that under our assumptions

$$E\left[e^{-(\lambda, X_{ki}^j(t-\theta_k^t))}|\mathcal{F}_{ki-1}(t)\right] = E\left[e^{-(\lambda, X_{ki}^j(t-\theta_k^t))}|\Re_0\right] = F_j(t-\theta_k^t, \mathbf{e}^{-\lambda}).$$

According to that lemma, in order to be

$$E\left[e^{-(\lambda_t, \mathbf{W}^*(t))}|\Re_0\right] \xrightarrow{P} H(\lambda) \tag{31}$$

with $\lambda_t = \lambda \oplus \mathbf{Q}(t)$ for any $\lambda \in R_+^n$, it is sufficient that

$$D(t,\lambda) \xrightarrow{P} H(\lambda), t \to \infty, \tag{32}$$

where

$$D(t,\lambda) = \prod_{k=1}^{M_t}\prod_{j=1}^{n}\prod_{i=1}^{N_t}\left\{F_j(t-\theta_k^t, \mathbf{e}^{-\lambda_t})\right\}^{\chi(i \le Y_{kj}^t)}$$

Since

$$D(t,\lambda) = T_t H(t,\lambda) + (1 - T_t) D(t,\lambda),$$

we have from Lemma 1, (27) and the choice M_t, that (32) holds for any $\lambda \in R^n_+$. Thus, due to the dominated convergence theorem, it follows from (31) that the Laplace transform of $\mathbf{Q}(t) \oplus \mathbf{W}^*(t)$ tends as $tghtarrow\infty$ to $EH(t,\lambda)$, that is

$$\mathbf{Q}(t) \oplus \mathbf{W}^*(t) \xrightarrow{\mathcal{D}} \mathbf{W}, t \to \infty. \tag{33}$$

It is not difficult to see that the inequality

$$|P\{\chi\mathbf{X} \leq \mathbf{x}\} - P\{\mathbf{X} \leq \mathbf{x}\}| \leq P\{\chi = 0\} \tag{34}$$

is true for any random vector $\mathbf{X}$, indicator χ and $\mathbf{x} \in R^n$.

The assertion of the theorem follows from (30), (33), (34) and the choice of M_t and N_t. The theorem is proved.

It is not difficult to see that using arguments of the proof of this theorem one can obtain limit theorems for generalized multitype (Bellman-Harris or Crump-Mode-Jagers) models of branching processes with immigration, when the limit theorem holds for the corresponding process without immigration.

References

[1] Athreya K., Ney P.(1972) Branching Processes, Springer-Verlag.

[2] Badalbaev I.S., Rahimov I.U.(1993) Non-Homogeneous Currents of Branching Processes, Tashkent, "Fan".

[3] Beska M., Klopotowski A., Slominski L. (1982) Limit theorems for random sums of dependent d-dimensional random vectors, Z.Wahrscheinlich. verw.Gebiete, 61, 43-57.

[4] Feller, W. (1968) An Introduction to Probability Theory and its Applications, V. II, J Wiley & Sons, New York.

[5] Jagers P. (1975) Branching Processes with Biological Applications, J. Wiley & Sons XIII, London.

[6] Kaplan N.(1974) Multidimensional age-dependent branching processes: the limiting distribution, J. Appl. Probab.,11, No2,225-236.

[7] Mode C.J.(1971) Multitype Branching Processes:theory and applications, Amer. Elsevier, XX, New York.

[8] Quine M.P.(1970) The multitype Galton-Watson process with immigration, J.Appl. Probab., 7, No2, 411-422.

[9] Rahimov I. (1992) General Branching Stochastic Processes with reproduction-dependent immigration, Theory Probab. and Appl. 37, No2, 513-525.

[10] Rahimov I.(1995) Random Sums and Branching Stochastic Processes, Springer-Verlag, Ser. LNS, 96, New York.

[11] Resnick S.I.(1987) Extreme Values, Regular Variation and Point Processes, Springer-Verlag, Ser.Apll.Probab.,4, New-York.

[12] Sevast'yanov B.A.(1971) Branching Processes, Nauka, Moscow.

[13] Shurenkov V.M.(1976) Two limit theorems for critical processes, Theory Probab. and Appl., 21, No3, 548-558.

[14] Vatutin V.A.,Zubkov A.M.(1993) Branching Processes.II, Journal of Soviet Mathematics, 67, No6, 3407-3485.

[15] Vatutin V.A.(1977) Limit theorems for critical Markov branching processes with several types and infinite second moments, Mathematics of the USSR - Sbornik, V.32, No.2, 215-225.

Applied Statistical Science, II
ISBN 1-56072-469-2

THE GENERALIZED ORDER STATISTICS

M. Ahsanullah
Department of Management Sciences
Rider University Lawrenceville, NJ 08648-3099, USA

1. INTRODUCTION

Recently Kamps (1995) introduced the generalized order statistics. The order statistics, record values and sequential order statistics are special cases of this generalised order statistics Suppose X(1,n,m,k), ..., X(n,n,m,k), ($k \geq 1$, m is a real number), are n generalized order statistics. Then the joint pdf $f_{1,\ldots,n}(x_1,\ldots,x_n)$ can be written as, (see Kamps (1995), pp 50-51),

$$f_{1,\ldots,n}(x_1,\ldots,x_n) = k \prod_{j=1}^{n-1} \gamma_j \prod_{i=1}^{n-1} (1-F(x_i))^m f(x_i)(1-F(x_n))^{k-1} f(x_n),$$

$$\text{for } F^{-1}(0) < x_1 < \cdots < x_n < F^{-1}(1),$$

$$= 0, \text{ otherwise,} \tag{1.1}$$

where $\gamma_j = k + (n-j)(m+1)$ and $f(x) = \dfrac{dF(x)}{dx}$.

For various interesting distributional properties of the generalized order statistics. see Kamps(1995).

If m = 0 and k = 1, then X(r,n,m,k) reduces to the ordinary rth order statistic and (1.1) is the joint pdf of the n order statistics $X_{1,n} \leq \ldots \leq X_{n,n}$. If k = 1 and m = -1, then (1.1) is the joint pdf of the first n upper record values of the independent and identically distributed random variables with distribution function F(x) and the corresponding probability density function f(x). for various distributional properties of record values see Ahsanullah (1995 a). In this paper we will consider the generalized order statistics of the two parameter exponential and uniform distribution.

2 . EXPONENTIAL DISTRIBUTION

Let X be a random variable (r.v.) whose probability density function (pdf) f is given by

$$f(x,\theta) = \sigma^{-1}\exp(-\sigma^{-1}(x-\mu)), \text{ for } x>\mu,\ \sigma>0,$$
$$= 0, \text{ otherwise.} \qquad (2.1)$$

We will denote $X \in E(\mu,\sigma)$ if the pdf of X is of the form given by (2.1).

2.1 DISTRIBUTIONAL PROPERTIES OF X(R,M,N,K)

Lemma 2.1.

If $X_i \in E(\mu,\sigma)$, I=1,2,... and X's are independent, then $\gamma_1 X(1,n,m,t)\ \varepsilon\ E(\mu,\sigma)$. and

$$X(r,n,m,k) \underline{\underline{d}} \mu + \sigma\sum_{j=1}^{r}\frac{W_j}{\gamma_j}.$$

Proof:

Integrating out $x_1,\ldots, x_{r-1}, x_{r+1},\ldots$ and x_n from (1.1), we get the pdf $f_{r,n,m,k}$ of X(r,n,m,k) (for details, see Kamps (1994), p. 64) as

$$f_{r,n,m,k}(x) = \frac{c_{r-1}}{(r-1)!}(1-F(x))^{k+(n-r)(m+1)-1} f(x)\, g_m^{r-1}(F(x)), \qquad (2.1.1)$$

where $c_{r-1} = \prod_{j=1}^{r}\gamma_j$ and

$$g_m(x) = \frac{1}{m+1}(1-(1-x)^{m+1}),\ m \neq -1,$$
$$= -\ln(1-x),\ m=-1,\ x\ \varepsilon\ (0,1).$$

Since $\lim_{m\to -1}\frac{1}{m+1}(1-(1-x)^{m+1}) = -\ln(1-x)$, we will write

$g_m(x) = \frac{1}{m+1}(1-(1-x)^{m+1})$, for all $x\ \varepsilon\ (0,1)$ and for all m with $g_{-1}(x) = \lim_{m\to -1} g_m(x))$.

Substituting $F(x) = \int_{-\infty}^{x} f(x)dx$ and f(x) as given by (2.1), we obtain for r = 1,

$$f_{1,n,m,k}(x) = \gamma_1 \sigma^{-\gamma_1 (x-\mu)}, \quad x > \mu, \sigma > 0. \tag{2.1.2}$$

Thus γ_1 X(1,n,m,t) $\in$ E(μ,σ).

From (2.1.2), we obtain the moment generating function

$M_{r,n,m,k}(t)$ of X(r,n,m,k) when X ε E(μ,σ) as

$$M_{r,n,m,k}(t) = \int_{\mu}^{\infty} \frac{e^{it\mu}}{\mu} \cdot \frac{c_{r-1}}{(r-1)!} e^{-\gamma_r \sigma^{-1}(x-\mu)} \left\{ \frac{1}{m+1}\left[1 - e^{-(m-1)\sigma^{-1}(x-\mu)}\right]\right\}^{r-1} dx$$

$$= \frac{c_{r-1}}{(r-1)!} e^{it\mu} \int_{0}^{\infty} e^{-y(\gamma_r - it\sigma)} \left\{\frac{1}{m+1}\left[1 - e^{-(m+1)y}\right]\right\}^{r-1} dy. \tag{2.1.3}$$

Using the property (see Gradsheyn and Ryzhik (1965), p. 305)

$$\int_{0}^{\infty} e^{-ay}(1 - e^{-by})^{r-1} dy = \frac{1}{b} B(\frac{a}{b}, r) \text{ and } \gamma_r + i(m+1) = \gamma_{r-i}$$

we obtain from (2.1.3)

$$M_{r,n,m,k}(t) = \frac{c_{r-1}}{(r-1)!} e^{i\mu t} \frac{(r-1)!}{\prod_{j=1}^{r} \gamma_j \left(1 - \frac{it\sigma}{\gamma_j}\right)} = e^{i\mu t} \prod_{j=1}^{r} \left\{1 - \frac{it\sigma}{\gamma_j}\right\}^{-1}. \tag{2.1.4}$$

Thus

$$X(r,n,m,k) \underline{\underline{d}}\, \mu + \sigma \sum_{j=1}^{r} \frac{W_j}{\gamma_j}, \tag{2.1.5}$$

where $\underline{\underline{d}}$ denotes equal in distribution and $W_1,\ldots,W_r$ are i.i.d. with $W_i \in E(0,1)$.

If we take k = 1,m = 0, then we obtain from (2.1.4) the well known result for order statistics for X $\in$ E(μ,σ)

$$X_{rn} = \mu + \sigma \sum_{j=1}^{r} W_j / (n-j+1) \tag{2.1.6}$$

From (2.1.5) if follows that γ_r {X(r,n,m,k) - X(r-1,n,m,k)} $\in$ E(0,σ). This property can also be obtained by considering the joint pdf of X(r,n,m,k), X(r-1,n,m,k) and using the transformation U_1 = X(r-1,n,m,k) and $W_r = \gamma_r$(X(r,n,m,k) - X(r-1,n,m,k)).

2.2 ESTIMATION OF μ AND σ

MINIMUM VARIANCE UNBIASED ESTIMATORS

Lemma 2.2

$\hat{\mu}$ and $\hat{\sigma}$ be the minimum variance unbiased estimates of μ and σ respectively based on X(1,n,m,k),..., X(n,n,m,k), then

$$\hat{\mu} = X(1,n,m,k) - \frac{\hat{\sigma}}{\gamma_1} \text{ and}$$

$$\hat{\sigma} = \frac{1}{n-1}\left[\sum_{j=1}^{n} (\gamma_j - \gamma_{j+1}) X(j,n,m,k) - \gamma_1 X(1,n,m,k)\right]$$

with $\gamma_{n+1} = 0$, $\text{Var}(\hat{\mu}) = \frac{n\sigma^2}{(n-1)\gamma_1^2}$, $\text{Var}(\hat{\sigma}) = \frac{\sigma^2}{n-1}$ and $\text{Cov}(\hat{\mu},\hat{\sigma}) = -\frac{\sigma^2}{(n-1)\gamma_1}$

Proof.

Using (2.4), it can be shown easily that

$$E(X(r,n,m,k)) = \mu + \alpha_r \sigma$$

$$\text{Var } X(r,n,m,k)) = \sigma^2 V_r, \text{ for } 1 \le r < s \le n,$$

where $\alpha_r = \sum_{j=1}^{r} \frac{1}{\gamma_j}$, $V_r = \sum_{j=1}^{r} \frac{1}{\gamma_j^2}$.

Let $X' = (X(1,n,m,k), X(2,n,m,k),..., X(n,n,m,k))$, then $E(x) = \mu\, 1 + \sigma\, \alpha$, $\text{Var}(x) = \sigma^2 V$, where 1 is a nx1 vector of units, $\alpha' = (\alpha_1, \alpha_2, ..., \alpha_n)$,

$V = (V_{ij})$ and $V_{ij} = V_i$ for $1 \le j \le n$.

Let $\Omega = V^{-1} = V^{ij}$, then

$$V^{ii} = \gamma_i^2 + \gamma_{i+1}^2, \; i = 1,2,...,kn, \; \gamma_{n+1} = 0$$

$$V^{i+1,i} = V^{i,i+1} = -\gamma_{i+1}^2$$

$$V^{ij} = 0 \text{ for } |i-j| > 1.$$

The minimum variance linear unbaised estimates $\hat{\mu}$, $\hat{\sigma}$ of μ and σ respectively are (see David (1981)).

$$\hat{\mu} = -\alpha' V^{-1}(1\alpha' - \alpha 1')V^{-1}X / \Delta$$

$$\hat{\sigma} = 1' V^{-1}(1\alpha' - \alpha 1')V^{-1}X / \Delta, \text{ where } \Delta = (1'V^{-1}1)(\alpha' V^{-1}\alpha) - (1'V^{-1}\alpha)^2,$$

with $\mathrm{Var}(\hat{\mu}) = \sigma^2 \alpha' V^{-1}\alpha / \Delta$, $\mathrm{Var}(\hat{\sigma}) = \sigma 1' V^{-1} 1 / \Delta$ and $\mathrm{Cov}(\hat{\mu}, \hat{\sigma}) = \sigma^2 1' V^{-1} \alpha / \Delta$.

It can be shown easily that

$$1'V^{-1} = \left(\gamma_1^2 \ 0 \ 0 \ \dots \ 0\right)$$

$$\alpha' V^{-1} = (\gamma_1 - \gamma_2 \ \ \gamma_2 - \gamma_3 \ \ \gamma_3 - \gamma_4 \cdots \gamma_{n-1} - \gamma_n \ \ \gamma_n)$$

$$\alpha' V^{-1}\alpha = n, \ 1'V^{-1}1 = \gamma_1^2, \ 1'V^{-1}\alpha = \gamma_1 \text{ and } \Delta = (n-1)\gamma_1^2 .$$

Now

$$1'V^{-1}(1\alpha' - \alpha 1)V^{-1} X/\Delta$$

$$= \frac{1}{\Delta}(1'V^{-1}1\alpha' - 1'V^{-1}\alpha 1')V^{-1}X$$

$$= \frac{1}{\Delta}(\gamma_1^2 \alpha' V^{-1}X - \gamma_1 1'V^{-1}X$$

$$= \frac{1}{n-1}(\alpha' V^{-1}X - \gamma_1 X(1,n,m,k))$$

Hence

$$\hat{\sigma} = \frac{1}{n-1}\left[\sum_{j-1}^{n} \left(\gamma_j - \gamma_{j+1}\right) X(j,n,m,k) - \gamma_1 X(1,n,m,k)\right].$$

We can write $\alpha' = \frac{1'}{\gamma_1} + c'$, where

$$c' = \left(0 \ \ \frac{1}{\gamma_2} \ \ \frac{1}{\gamma_2} + \frac{1}{\gamma_3} \ \ \dots \frac{1}{\gamma_2} + \frac{1}{\gamma_3} + \dots + \frac{1}{\gamma_n}\right)$$

Thus

$$\hat{\mu} = -c^1 V^{-1}(1\alpha' - \alpha 1')V^{-1} X/\Delta - \frac{\hat{\sigma}}{\gamma_1}.$$

We have

$$c' V^{-1} 1 = 0, \ c'V^{-1}\alpha = n-1 \text{ and hence } \hat{\mu} = X(1,r,m,k) - \frac{\hat{\sigma}}{\gamma_1}.$$

If $m = 0$ and $k = 1$, then $\gamma_j = n-j+1$ and $\hat{\mu}$ and $\hat{\sigma}$ coincide with the MVLUEs given by the order statistics (see Arnold *et al* (1992) p. 176).

If $m= -1$ and $k =1$, then $\gamma_j = 1$ and $\hat{\mu}$ and $\hat{\sigma}$ coincide with the MVLUEs estimates given by Ahsanullah (1980 p. 366).

The variances and the covariance of $\hat{\mu}, \hat{\sigma}$ are

$$\operatorname{Var}(\hat{\mu}) = \frac{\sigma^2 \alpha' V^{-1} \alpha}{\Delta} = \frac{n\sigma^2}{(n-1)\gamma_1^2}$$

$$\operatorname{Var}(\hat{\sigma}) = \frac{\sigma^2 \alpha' V^{-1} 1}{\Delta} = \frac{\sigma^2}{n-1}$$

$$\operatorname{Cov}(\hat{\mu}, \hat{\sigma}) = -\frac{\sigma^2 \alpha' V^{-1} \alpha}{\Delta} = \frac{\sigma^2}{(n-1)\gamma}.$$

Best Linear Invariant Estimators

Lemma 2.2.2

The best linear invariant (in the sense of minimum mean squared error and invariance with respect to the location parameter μ) estimators (BLIE) $\tilde{\mu}$ $\tilde{\sigma}$ of μ and σ are

$$\tilde{\mu} = \hat{\mu} - \hat{\sigma}\left(\frac{E_{12}}{1+E_{22}}\right) \quad \text{and} \quad \tilde{\sigma} = \hat{\sigma} / (1 + E_{22}),$$

where $\hat{\mu}$ and $\hat{\sigma}$ are MVLUE of μ and σ and

$$\begin{pmatrix} \operatorname{Var}(\hat{\mu}) & \operatorname{Cov}(\hat{\mu}, \hat{\sigma}) \\ \operatorname{Cov}(\hat{\mu}, \hat{\sigma}) & \operatorname{Var}(\hat{\sigma}) \end{pmatrix} = \sigma^2 \begin{pmatrix} E_{11} & E_{12} \\ E_{12} & E_{22} \end{pmatrix}.$$

The mean squared errors of these estimators are

$$\operatorname{MSE}(\hat{\mu}) = \sigma^2 \left\{ E_{11} - E_{12}^2 (1 + E_{22})^{-1} \right\}$$

and

$$MSE(\tilde{\sigma}) = \sigma^2 E_{22}(1+E_{22})^{-1}.$$

Substituting the values of E_{12} and E_{22} in the above equations, we get on simplification

$$\tilde{\mu} = \hat{\mu} + \frac{1}{n\gamma_1}\hat{\sigma} \quad \text{and} \quad \tilde{\sigma} = \frac{n-1}{n}\hat{\sigma}$$

$$MSE(\tilde{\mu}) = \frac{n+1}{n}\frac{\sigma^2}{\gamma_1^2}, \quad \text{and} \quad MSE(\tilde{\sigma}) = \frac{1}{n}\sigma^2 .$$

2.3. PREDICTION OF X(s,n,m,k)

We shall assume $s > n$. Let $\eta' = (\eta_1, \eta_2, \ldots, \eta_n)$ where $\eta_j = Cov(X(s,n,m,k), X(j,n,m,k))$, $j = 1,2,..,n$ and $\alpha^* = \sigma^{-1}E(X(x,n,m,k) - \mu)$. The best linear unbiased predictor (BLUP) $\hat{X}(x,n,m,k)$ of $X(s,n,m,k)$ is $\hat{X}(x,n,m,k) = \hat{\mu} + \alpha^*\hat{\sigma} + \eta V^{-1}(X - \hat{\mu}\, 1 - \hat{\sigma}\, \alpha)$, where $\hat{\mu}$ and $\hat{\sigma}$ are the MVLE of μ and σ respectively. But $\alpha^* = \alpha_s$ and $\eta' = (V_1, V_2, \ldots, V_n)$. It can be shown $\eta' V^{-1} = (0,0,\ldots,1)$ and hence

$$\begin{aligned}\hat{X}(s,n,m,k) &= \hat{\mu} + \alpha_s\, \hat{\sigma} + X(n,n,m,k) - \hat{\mu} - \alpha_n\, \hat{\sigma} \\ &= X(n,n,m,k) + (\alpha_s - \alpha_n)\, \hat{\sigma} \qquad (2.3.1)\end{aligned}$$

If $m = 0$ and $k = 1$, then $\gamma_j = n-j+1$ and $\hat{X}(s,n,m,k)$ coincide with the BLUP based on the order statistics (see Arnold *et al* (1992) p. 181).

If $m = -1$ and $k = 1$, then $\gamma_j = 1$ and $\hat{X}(s,n,m,k)$ coincide with the BLUP based on record values (see Ahsanullah (1980 p. 367).

We have $E(\hat{X}(x,n,m,k) = \mu + (\alpha_s - \alpha_n)\sigma = \mu + \alpha_s\sigma$

$$Var(\hat{X}(x,n,m,k)) = \sigma^2 V_n + (\alpha_s - \alpha_n)^2 \frac{\sigma^2}{n-1} + 2(\alpha_s - \alpha_n)Cov(X(n,n,m,k), \hat{\sigma})$$

$$= \sigma^2 \left\{ \left(\frac{1}{\gamma_1^2} + .. + \frac{1}{\gamma_n^2} \right) + \frac{1}{n-1}\left(\frac{1}{\gamma_{n+1}} + .. - \frac{1}{\gamma_s} \right)^2 + \frac{2}{n-1}\left(\frac{1}{\gamma_{n+1}} + .. + \frac{1}{\gamma_s} \right)\left(\frac{1}{\gamma_2} + .. + \frac{1}{\gamma_n} \right) \right. . \qquad (2.3.2)$$

$$\begin{aligned}MSE(\hat{X}(x,n,m,k) &= E(\hat{X}(x,n,m,k) - X(s,n,m,k))^2 \\ &= E[X(n,n,m,k) - X(x,n,m,k) + (\alpha_s - \alpha_n)\hat{\sigma}\,]^2\end{aligned}$$

$$= \sigma^2(V_n + V_s - 2V_n + (\alpha_s - \alpha_n)^2 \frac{1}{n-1})$$

$$= \sigma^2(V_s - V_n + \frac{(\alpha_s - \alpha_n)^2}{n-1})$$

If we take k = 1, m = -1, then the BLUP $\hat{X}_{U(s)}$ of the sth upper record value from (2.3.1) is

$$\hat{X}_{U(s)} = \frac{(x-1)X_{u(n)} - (s-n)X_{u(1)}}{n-1} \tag{2.3.3}$$

and from (2.3.2)

$$E(\hat{X}_{U(s)}) = \sigma^2(m+s^2-2s)/(m-1) \tag{2.3.4}$$

The equations (2.3.3) and (2.3.4) were obtained by Ahsanullah (1980).

Let $\tilde{X}$(x,n,m,k) be the best linear invariant predictor of X(s,n,m,k). Then

$$\tilde{X}(s,n,m,k) = \hat{X}(s,n,m,k) - \frac{c_{12}}{1+c_{22}}\hat{\sigma}, \tag{2.3.5}$$

where

$$c_{12}^{*}\sigma^2 = \mathrm{Cov}(\hat{\sigma}, (1-\eta' V^{-1} 1)\hat{\mu} + (\alpha^* - \eta V^{-1}\alpha)\hat{\sigma}) \text{ and } c_{22}\sigma^2 = \mathrm{Var}(\hat{\sigma}).$$

It can easily be shown that $c_{12}^{*} = \frac{\alpha_s - \alpha_n}{n-1}$. Since $c_{22} = \frac{1}{n-1}$, we have $\frac{c_{12}^{*}}{1+c_{22}} = \frac{\alpha_s - \alpha_n}{n}$. Thus

$$\tilde{X}(s,n,m,x) = \hat{X}(s,n,m,k) - \frac{\alpha_s - \alpha_n}{n}\hat{\sigma}$$

$$= X(n,n,m,k) + \frac{n-1}{n}(\alpha_s - \alpha_n)\hat{\sigma} \tag{2.3.6}$$

$$E(\tilde{X}(s,n,m,k)) = \mu + (\alpha_s + \frac{\alpha_s - \alpha_n}{n})\sigma \tag{2.3.7}$$

and

$$\mathrm{Var}(\tilde{X}(x,n,m,k)) = \sigma^2\{V_n + \left(\frac{n-1}{n}\right)^2 (\alpha_s - \alpha_n)^2 \frac{1}{n-1}$$

$$= \sigma^2\left\{\frac{1}{\gamma_1^2}+..+\frac{1}{\gamma_n^2}+\frac{n-1}{n^2}\left(\frac{1}{\gamma_{n+1}}+..+\frac{1}{\gamma_s}\right)^2\right.$$

$$\left. + \ 2\left(\frac{n-1}{n}\right)\left(\frac{1}{\gamma_{n+1}}+..\frac{1}{\gamma_s}\right)\left(\frac{1}{\gamma_1}+..+\frac{1}{\gamma_n}\right)\right\} \qquad (2.3.8)$$

$$E(\tilde{X}(s,n,m,k)) \ = E(\tilde{X}(s,n,m,k) - X(s,n,m,k)^2$$

$$= E(X(n,n,m,k) - X(s,n,m,k) + \frac{n-1}{n}(\alpha_s - \alpha_n)\hat{\sigma})^2$$

$$= \sigma^2\left[\gamma_n + \gamma_s - 2\gamma_n + \frac{n-1}{n^2}(\alpha_s - \alpha_n)\right]$$

$$= \sigma^2\left[\gamma_s + \gamma_n + \frac{n-1}{n^2}(\alpha_s - \alpha_n)\right]$$

$$E(\tilde{X}(s,n,m,k) - X(s,n,m,k) = (\alpha_n - \alpha_s)\sigma + \frac{n-1}{n}(\alpha_s - \alpha_n)\sigma$$

$$= - \frac{1}{n}(\alpha_s - \alpha_n)\sigma$$

Thus $\text{MSE}\ (\tilde{X}(s,n,m,k))$

$$= E(\tilde{X}(s,n,m,k) - X(s,n,m,k))^2 + \frac{1}{n^2}(\alpha_s - \alpha_n)\sigma^2$$

$$= \sigma^2\left[\gamma_s - \gamma_n + \frac{1}{n}(\alpha_s - \alpha_n)^2\right]$$

$$= \text{MSE}(\hat{X}(s,n,m,k)) - \frac{(\alpha_s - \alpha_n)^2}{n(n-1)}$$

For k = 1, m = -1, the equations (3.7) and (3.8) become

$$E(\hat{X}_{U(s)}) = \mu + (s + \frac{s-n}{n})\sigma$$

$$\text{Var}(\hat{X}_{U(s)}) = \sigma^2\ \frac{n^2 + ns^2 - s^2}{n^2}$$

which were given by Ahsanullah (1980).

2.4 Characterizations

Theorem 2.4.1

Let X be a non-negative r.v. having an absolutely continuous (with respect to Lebesgue measure) strictly increasing distribution function F(x) for all x > 0 and F(x) < 1 for all x > 0. Then the following properties are equivalent

(a) X has an exponential distribution with density as given in (1.1);

(b) X has a monotone hazard rate and for one k, one r, one n and one m, the statistics {k +(n-r)(m+1)} {X(r,n,m,k) - X(r-1,n,m,k)} and X are identically distributed.

Proof:

Integrating out $x_1,\ldots, x_{r-2},\ldots, x_n$, we get from the joint pdf $f_{r-1,r}(x,y)$ of X(r-1,n,m,k) and X(r,n,m,k), $1 < r \le n$, as

$$f_{r-1,r}(x,y) = \frac{c_{r-2}}{(r-2)!}(1-F(x))^m g_m^{r-2}(F(x))\ (1-F(y))^{\gamma_r - 1}\ f(x)\ f(y)$$

$$\text{for } F^{-1}(0)<x< y <F^{-1}(1) \qquad (2.4.1)$$

where $c_{r-2} = \prod_{j=1}^{r-1} \gamma_j,\ \gamma_j = k + (n-j)(m+1),\ j = 1,2,..,n$ and

$$g_m(x) = \frac{1}{m+1}\left[1-(1-F(x))^{m+1}\right],\quad m \ne -1$$

$$= -\ln(1-F(x)), \qquad m = -1$$

Using the transformation U = X(r-1,n,m,k) and W = γ_r (X(r,n,m,k) - X(r-1,n,m,k)), we get on simplification the joint pdf $f_{UW}(u,w)$ of U and W as

$$f_{UW}(u,w) = \frac{c_{r-2}}{(r-2)!}(1-F(u))^m g_m^{r-2}(F(u))(1-F(u+\frac{w}{\gamma_r}))^{\gamma_r - 1} f(u)f(u+\frac{w}{\gamma_r}),$$

$$= 0, \text{ otherwise.} \qquad (2.4.2)$$

If X has the pdf as given in (2.1), then 1 - F(x) = $e^{-\theta x}$ and

$$g_m(F(x)) = \frac{1}{m+1}\left(1-e^{-(m+1)\theta x}\right),\ m \ne -1$$

$$= \theta x \qquad ,\ m = -1.$$

On simplification, we get from (2.4.1)

$$f_{UW}(u,w) = \frac{c_{r-2}\theta^2}{(r-2)!}\; g_m^{r-2}(1-e^{-\theta u})\; e^{-\gamma_{r-1}\theta u}\; e^{-\theta w} \quad 0 < u, w < \infty$$

$$= 0, \quad \text{otherwise.}$$

Thus W and X are identically distribution.

Suppose X and W are identically distributed, then from (2.4.2) we must have

$$f(w) = \int_0^\infty \frac{c_{r-2}}{(r-2)!}(1-F(u))^m g_m^{r-2}(F(u))\left(1-F\left(u+\frac{w}{\gamma_r}\right)\right)^{\gamma_r - 1} \cdot f(u) f\left(u+\frac{w}{\gamma_r}\right) du \qquad (2.4.3)$$

Integrating both sides of (2.4.3) with respect to w from w_1 to ∞ and using the equations

$$1 = \int_0^\infty \frac{c_{r-2}}{(r-2)!}(1-F(u))^{-1+\gamma_{r-1}} g_m^{r-2}(F(u)) f(u) du,$$

and $\gamma_{r-1} = \gamma_r + m + 1$, we get

$$\int_0^\infty \frac{c_{r-2}}{(r-2)!}(\bar{F}(u))^{-1+\gamma_{r-1}} g_m^{r-2}(F(u))\; H(u, w_1)\, du = 0, \text{for all } w_1, \quad 0 < w_1 < \infty, \text{ where}$$

$$H(u,w) = \bar{F}(w) - \left(\frac{\bar{F}(u + \frac{w}{\gamma_r})}{\bar{F}(u)}\right)^{\gamma_r}, \quad \bar{F}(x) = 1 - F(x) \qquad (2.4.4)$$

By Lemma 5.1 given in the Appendix, we have $H(0,w_1) = 0$, for all $w_1 > 0$, i.e.

$$\bar{F}(w) = \left(\bar{F}\left(\frac{w}{\gamma_r}\right)\right)^{\gamma_r}$$

for all $w > 0$. Substituting $\psi(w) = -\ln \bar{F}(w)$, we get

$$\psi(w) = \gamma_r\, \psi\left(\frac{w}{\gamma_r}\right) \qquad (2.4.5)$$

for all w and some fixed γ_r. We know that F(x) is strictly monotone increasing with F(x) < 1 for all x > 0. Hence $\psi(x)$ is continuous for all x > 0. Thus the solution of (2.4.5) (see Aczel (1966), p. 31) is

$$\psi(x) = \theta x, \qquad (2.4.6)$$

where θ is a constant. Using the boundary condition $F(0) = 0$ and $F(\infty) = 1$, we get,

$$F(x) = 1 - e^{-\theta x}, \text{ where } \theta > 0.$$

Remark 1:

If we take k = 1, m = 0, then the Theorem 2.1 gives a characterization of the exponential distribution based on the identical distribution of $(n-r+1)(X_{r,n} - X_{r-1,n})$ and X (Ahsanullah (1977)).

Remark 2:

If we take k=1, m= -1, then the theorem 2.1 gives a characterization of the exponential distribution based on the identical distribution of $X_{u(m)} - X_{u(m-1)}$ ($X_{u(m)}$ is the nth upper record value and X (see Ahsanullah (1978)).

A. APPENDIX

Lemma A.1.

Let F(x) be the absolutely continuous (with respect to Lebesgue measure) monotone strictly increasing distribution function of a non-negative r.v. X with F(0) = 1, F(x) < 1 for all x.. Assume f(x) as the corresponding pdf. If the hazard rate is monotone and if

$$\int_0^\infty \frac{c_{r-2}}{(r-2)!}(\overline{F}(u))^{-1+\gamma_{r-1}}\, g_m^{r-2}(F(u))\ H(u,w)\, du = 0, \qquad \text{(A.1)}$$

for all w, one r, one m, and one γ_{r-1},

where $H(u,w) = \overline{F}(w) - \left(\frac{\overline{F}(u+\frac{w}{\gamma_r})}{\overline{F}(u)}\right)^{\gamma_r}$, $\overline{F}(x) = 1 - F(x)$,

then $H(0,w) = 0$ for all $w > 0$.

Proof:

We have $H(u,0) = 0 = H(u,\infty)$.

$$\frac{\partial H(u,w)}{\partial u} = q(u,w)(h(u+\frac{w}{\gamma_r}) - h(u)), \qquad \text{(A.2)}$$

where $q(u,w) = \left(\frac{\overline{F}(u+\frac{w}{\gamma_r})}{\overline{F}(u)}\right)^{\gamma_r}$,

(i) If h(x) = constant for all x, then $\frac{\partial H(u,w)}{\partial u}$=0, for all u and w.

(ii) If h(x) is strictly monotone increasing in x, then it follows from (A.2) that H(u,w) is strictly increasing in u for any fixed w. Thus for all w, if H(u,w) ≢ 0, then H(u,w) must be negative in an interval including zero. Let u ε I, where I in the interval [0,b], b is a real number and H(u,w) < 0 for u ε I. Now for u ε I,

$$\frac{\partial H(u,w)}{\partial w} = -f(w) + \left(\frac{\overline{F}(u+\frac{w}{\gamma_r})}{\overline{F}(u)} \right)^{\gamma_r} h(u+\frac{w}{\gamma_r})$$

$$> - f(w) + \overline{F}(w)\ h(u + \frac{w}{\gamma_r}) \text{ for } u \ \varepsilon\ I, \text{ since } H(u,w) < 0$$

$$> \ \overline{F}(w)\,[\,h(u + \frac{w}{\gamma_r}) - h(w)] < 0,$$

for all u ε I and all $w > \frac{\gamma_r}{\gamma_r - 1} u$. Thus for all u ε I, H(u,w) decreases to 0 as w → ∞.

But for all u ε I, H(u,0) = 0, H(u,w) < 0 and H(u,w) decreases to 0 as w → ∞. Therefore it follows by the continuity of H(u,w) that H(u,w) = 0 for all w > 0 and for u ε I. Hence H(0,w) = 0 for all w > 0..

(iii) If h(x) is strictly monotone decreasing in x, then following a similar argument to that in (ii) it can be shown that G(0,w) = 0 for all w > 0.

3.1 Generalized order statistics from two parameter uniform distribution

Let X be a random variable (r.v.) whose probability density function (pdf) f is given by

$$f(x,\mu,\sigma) \qquad = \frac{1}{\sigma}, \quad -\infty < \mu < x < \mu + \sigma < \infty, \ \sigma > 0$$

$$= 0, \ \text{otherwise.} \qquad (3.1)$$

We will denote $X \in U(\mu , \sigma)$ if the pdf of X is of the form given by (3.1).

David (1981) studied the minimum variance unbiased estimators of μ and σ based on n order statistics. Ahsanullah (1986) presented the estimators of μ and σ based on first n upper values. In this paper we will consider the estimators of μ and σ based on the first n generalized order statistics. The results of David (1981) and Ahsanullah (1986) are obtained as special cases of the results of this paper.

Lemma 3.1

$E(X(r,n,m,k)) = \mu+\sigma-\sigma c_r/d_r$, $Var(X(r,n,m,k) = \sigma^2\left(c_r/e_r-(c_r/d_r)^2\right)$

and for $s > r$, $Cov(X(s,n,m,k), X(r,n,m,k)) = a_r b_s \sigma^2$, where

$a_r = (d_r/c_r)\left(c_r/e_r-(c_r/d_r)^2\right)$, $b_r = c_r/d_r$ $c_r = \prod_{j=1}^{r}\gamma_j$, $d_r = \prod_{j=1}^{r}(\gamma_j+1)$ and

$e_r = \prod_{j=1}^{r}(\gamma_j+2)$, $r = 1,\ldots,n$.

Proof:

For $X \in U(\mu, \sigma)$, we have

$$f_{r,n,m,k}(x) = \frac{c_r}{(r-1)!\sigma}\left(\frac{\mu+\sigma-x}{\sigma}\right)^{\gamma_r-1}\left(1-\left(\frac{\mu+\sigma-x}{\sigma}\right)^{m+1}\right)^{r-1}(m+1)^{-(r-1)} \tag{3.2}$$

Using the equation (see Gradshteyn and Ryzhik (1980) p. 294) $\lambda\int_0^1 x^{\mu-1}(1-x^{\lambda})^{\nu-1}dx =$ $B(\frac{\mu}{\lambda},\nu)$ and the transformation $Y_r = \frac{\mu+\sigma-X(r,n,m,k)}{\sigma}$, we obtain on simplification from (6.2), $E(Y) = c_r/d_r$ and $E(Y^2) = c_r/e_r$. $E(Y)$ and $E(Y^2)$ can be obtained as special cases from the moments of the generalized order statistics of power function distribution (see Kamps (1995) p.99). Thus $E(X(r,n,m,k)) = \mu+\alpha_r\sigma$, $\alpha_r = 1 - b_r$, $b_r = c_r/d_r$ and $Var(X(r,n,m,k)) = \sigma^2\left(c_r/e_r-(c_r/d_r)^2\right)$. The conditional pdf $f(y|x)$ of $X(s,n,m,k)$, given $X(r,n,m,k) = x$ for $s > r$, is given by

$$f_c(y|x) = \frac{c_s}{c_r}\frac{1}{(s-r-1)!}\frac{1}{(m+1)^{s-r-1}}\left[1-\left((\mu+\sigma-y)/(\mu+\sigma-x)\right)^{m+1}\right]^{s-r-1} \cdot\left(\frac{\mu+\sigma-y}{\mu+\sigma-x}\right)^{\gamma_s-1}\frac{1}{\mu+\sigma-x}, \text{for } x<y<\mu+\sigma. \tag{3.3}$$

Using the transformation $Z_r = (\mu+\sigma - X(s,n,m,k))(\mu+\sigma - X(r,n,m,k))^{-1}$, we obtain from (3.3) $E(Z_r \mid X(r,n,m,k) = x) = (c_s/c_r)(d_r/d_s)$. Thus $E(X(r,n.m,k) = \mu+\sigma -(\mu+\sigma - x)$ $(c_s/c_r)(d_r/d_s)$ and for $s > r$, Cov(X(s,m,n,k), X(r,n,m,k)) = E(X(s,r,n,k) (X(r,n,m,k)- E(X(r,n,m,k)) $= (c_s/c_r)(d_r/d_s)$ E((X(r,n,m,k ((X(r,n,m,k) - E(X(r,n,m,k)) = $(c_s/c_r)(d_r/d_s)\ \mathrm{Var}(X(r,n,m,k))$. Writing $\sigma^2\ V_{rs}$ = Cov (X(s,n,m,k), X(r,n,m,k)) for $s \geq r$, we have $V_{rs} = a_r\, b_s$ where $a_r = (d_r/c_r)\left(c_r/e_r - (c_r/d_r)^2\right)$, $b_r = c_r/d_r$, $r = 1, \ldots n$

3.1 Estimation of μ and σ

Minimum variance linear unbiased estimators

Let X(1,n,m,k), ..., X(n,n,m,k) be n generalized order statistics from $U(\mu, \sigma)$.

Theorem 3.1.1

The minimum variance linear unbiased estimators (MVLUE) of $\hat{\mu}$, $\hat{\sigma}$ of μ and σ are given by $\hat{\sigma} = (X(1,n,m,k) - \hat{\mu})(\gamma_1 + 1)$ and $\hat{\mu} = \sum_{i=1}^{n} w_{1i}\, X(i,n,m,k)$, where

$$w_{11} = \frac{1}{D_0}\frac{e_1}{c_1}\left[(\gamma_1+2)(\gamma_2+2) + (\gamma_1+1)(T\gamma_1 - \gamma_1 - 2)\right],$$

$$w_{21} = \frac{(\gamma_1+1)e_1}{D_0 c_1}\left[(-T\gamma_1 + \gamma_1 + 2) - (\gamma_1+2)(\gamma_2+2)\right], w_{1n} = -(\gamma_1+2)(\gamma_n+1)\frac{e_n}{c_n D_0}$$

and $w_{2j} = -(\gamma_1 + 1) w_{1j}, j = 1,2,\ldots,n.$

Proof.

We write $E(X) = \mu\ 1 + \sigma\ \alpha$, where $1' = (1,1,\ldots,1)$, $\alpha' = (\alpha_1, \alpha_2, \ldots, \alpha_n)$ and $X' = (X_1, X_2, \ldots, X_n)$. Let $V = (V_{ij})$. Since V is positive definite, it can written as (see Graybill (1969, p. 268) $V = T'T$, where

$$T' = \begin{bmatrix} t_{11} & 0 & 0 & 0 & 0 \\ t_{12} & t_{22} & 0 & 0 & 0 \\ t_{13} & t_{23} & t_{33} & 0 & 0 \\ . & . & . & . & . \\ t_{1n} & t_{2n} & t_{3n} & . & t_{nn} \end{bmatrix} \text{ with } t_{11} = \sqrt{V_{11}},\ t_{1j} = \frac{V_{1j}}{t_{11}},\ j = 2,3,\ldots,n;$$

$t_{ii} = \sqrt{V_{ii} - \sum_{p=1}^{i-1} t_{pi}^2}$ $t_{ij} = \frac{1}{t_{ii}}\left[V_{ij} - \sum_{p=1}^{i-1} t_{pi} t_{pj}\right], j>i,$ $t_{ij} = 0, j< i$ and $i = 2,3,...,n.$

We use the transformation $Y = AX$, $Y' = (Y_1, Y_2, ..., Y_n)$, and $A = (T')^{-1}$. Substituting

$V_{rs} = a_r b_s$ with $a_r = \frac{d_r}{c_r}\left[\frac{c_r}{e_r} - \left(\frac{c_r}{d_r}\right)^2\right]$, $b_s = \frac{c_s}{d_s}$, $r,s = 1,2,...,n$, we obtain on simplification

A as a lower triangular matrix with

$$A = \begin{pmatrix} a_{11} & 0 & 0 & 0 & 0 \\ a_{21} & a_{22} & 0 & 0 & 0 \\ 0 & a_{32} & a_{33} & 0 & 0 \\ . & . & . & . & . \\ 0 & 0 & 0 & a_{nn-1} & a_{nn} \end{pmatrix}, \text{ where } a_{11} = (\gamma_1 + 1)\sqrt{\frac{\gamma_1 + 2}{\gamma_1}},$$

$$a_{k\,k-1} = -\gamma_k \sqrt{\frac{e_k}{c_k}} \text{ and } a_{kk} = (\gamma_k + 1)\sqrt{\frac{e_k}{c_k}}, \ k = 2,...,n.$$

We have $E(Y) = B\,\theta$, with $\theta' = (\mu, \sigma)$, $B' = \begin{pmatrix} b_{11} & b_{21} & \cdot & \cdot & b_{n1} \\ b_{12} & b_{22} & \cdot & \cdot & b_{n2} \end{pmatrix}$,

$b_{11} = (\gamma_1 + 1)\sqrt{\frac{e_1}{c_1}}$ $b_{12} = -\sqrt{\frac{e_1}{c_1}}$, $b_{k1} = \sqrt{\frac{e_k}{c_k}}$ and $b_{k2} = -\sqrt{\frac{e_k}{c_k}}$, $k = 2,3,...,n.$

The minimum variance linear unbiased estimators $\hat{\mu}$ and $\hat{\sigma}$ of μ and σ respectively are given by

$\hat{\theta} = \begin{pmatrix} \hat{\mu} \\ \hat{\sigma} \end{pmatrix} = (B'B)^{-1}B'AX$. Let $(B'B)^{-1}B'A = W = \begin{pmatrix} w_{11} & w_{12} & \cdot & \cdot & \cdot & w_{1n} \\ w_{21} & w_{22} & \cdot & \cdot & \cdot & w_{2n} \end{pmatrix}$, then

$$w_{11} = \frac{1}{D_0}\frac{e_1}{c_1}\left[(\gamma_1 + 2)(\gamma_2 + 2) + (\gamma_1 + 1)(T\gamma_1 - \gamma_1 - 2)\right], w_{1j} = -(\gamma_{j+1} - \gamma_j + 1)\frac{e_j(\gamma_1 + 1)}{D_0 c_j}$$

$$w_{21} = \frac{(\gamma_1 + 1)e_1}{D_0 c_1}\left[(-T\gamma_1 + \gamma_1 + 2) - (\gamma_1 + 2)(\gamma_2 + 2)\right], w_{1n} = -(\gamma_1 + 2)(\gamma_n + 1)\frac{e_n}{c_n D_0}$$

and $w_{2j} = -(\gamma_1 + 1)w_{1j}, j = 1,2,...,n.$

On simplification, we get

$\hat{\sigma} = (X(1,n,m,k) - \hat{\mu}))(\gamma_1 + 1)$ and

$$\hat{\mu} = \sum_{i=1}^{n} w_{1i} X(i,n,m,k). \tag{3.1.1}$$

Since $(B'B)^{-1} = \frac{1}{D_0}\begin{pmatrix} T & -(T+\gamma_1+2) \\ -(T+\gamma_1+2) & T+(\gamma_1+2)^2 \end{pmatrix}$, the variances and the covariance of the estimators are

$$Var(\hat{\mu})=\sigma^2\frac{T}{D_0},\ var(\hat{\sigma})=\sigma^2\frac{T+(\gamma_1+2)^2}{D_0} \text{ and } Cov(\hat{\mu},\hat{\sigma})=-\sigma^2\frac{(T+\gamma_1+2)}{D_0}. \qquad (3.2.2)$$

David (1981) considered the unbiased minimum variance unbiased estimators (UMVU) of μ and ω based on the n order statistics, $X_{1,n} < X_{2,n} < \ldots < X_{n,n}$, from $U((\mu-\frac{1}{2}\omega,\omega)$. He obtained the unique UMVU of μ and ω as M and W' where $M = \frac{1}{2}(X_{1,n}+X_{n,n})$ and $W'=\frac{n+1}{n-1}(X_{n,n}-X_{1,n})$.

For order statistics, we have k=1, m = 0 and $\gamma_j = n-j+1$ On simplification, we get

$$T=\frac{n+2}{n}+\frac{(n+2)(n+1)}{n(n-1)}+\frac{(n+2)(n+1)n}{n(n-1)(n-2).}+\ldots+\frac{(n+2)(n+1).....3}{n(n-1).......1}=n(n+2)$$

$$D_0=(n+2)^2(n^2-1), w_{11}=\frac{n}{n-1}, w_{1k}=0, k=2,3,\ldots,,n-1,\ w_{1n}=-\frac{1}{n-1}.$$

Thus we obtain the minimum variance linear unbiased estimators of μ and σ from $U(\mu, \sigma)$ from (3.1.1) based on the n order statistics, $X_{1,n} < X_{2,n} < \ldots < X_{n,n}$ as $\hat{\mu}=\frac{nX_{1n}-X_{nn}}{n-1}$ and $\hat{\sigma}=\frac{n+1}{n-1}(X_{nn}-X_{1n})$.

For the upper record values, we have k = 1, m = -1 and γ_j = 1, j = 1,2,.... Hence the MVLUE estimators $\hat{\mu}$, $\hat{\sigma}$ of μ and σ (see Ahsanullah (1986)) based on the upper record values $X_{U(1)}, \ldots, X_{U(n)}$, are $\hat{\sigma}=2(X_{U(1)}-\hat{\mu})$, and

$$\hat{\mu}=\frac{1}{D_0}\left[3^{n+2}X_{U(1)}+2\sum 3^{k+1}X_{U(k)}-2.3^{n+1}X_{U(n)}\right] \text{ and } D_0=\frac{27}{2}\left(3^{n-1}-1\right).$$

Best Linear Invariant Estimators

Theorem 3.1.2

The best linear invariant (in the sense of minimum mean squared error and invariance with respect to the location parameter μ) estimators (BLIE) $\tilde{\mu}$ $\tilde{\sigma}$ of μ and σ are

$$\tilde{\mu} = \hat{\mu} - \hat{\sigma}\left(\frac{E_{12}}{1+E_{22}}\right) \text{ and } \tilde{\sigma} = \hat{\sigma} / (1+E_{22}), \tag{3.1..3}$$

where $\hat{\mu}$ and $\hat{\sigma}$ are MVLUE of μ and σ and

$$\begin{pmatrix} Var(\hat{\mu}) & Cov(\hat{\mu},\hat{\sigma}) \\ Cov(\hat{\mu},\hat{\sigma}) & Var(\hat{\sigma}) \end{pmatrix} = \sigma^2 \begin{pmatrix} E_{11} & E_{12} \\ E_{12} & E_{22} \end{pmatrix}.$$

The mean squared errors of these estimators are

$$MSE(\hat{\mu}) = \sigma^2 \left\{ E_{11} - E_{12}^2 (1+E_{22})^{-1} \right\}$$

and

$$MSE(\tilde{\sigma}) = \sigma^2 E_{22} (1+E_{22})^{-1}. \tag{3.1.4}$$

Substituting the values of E_{12} and E_{22} in (3.1.4) we get

$$\tilde{\mu} = \hat{\mu} + \frac{T+\gamma_1+2}{T(1+\gamma_1(\gamma_1+2))}\hat{\sigma} \text{ and } \tilde{\sigma} = \hat{\sigma}\frac{(\gamma_1+2)(T\gamma_1-\gamma_1-2)}{T(1+\gamma_1(\gamma_1+2))} \tag{3.1.5}$$

For the order statistics, we obtain $\mu = \hat{\mu} + \frac{\hat{\sigma}}{n(n+1)}$, $\tilde{\sigma} = \frac{(n-1)(n+2)}{n(n+1)}\hat{\sigma}$ with

$$MSE(\tilde{\mu}) = \frac{\sigma^2}{n(n+2)} \text{ and } MSE(\tilde{\sigma}) = \frac{2\sigma^2}{n(n+1)}.$$

Substituting $\gamma_1 = 1$ and $T = \frac{3(3^n - 1)}{2}$ in (3.1.5), we can obtain the corresponding estimators $\tilde{\mu}$ and $\tilde{\sigma}$ based on the n upper record values, $X_{U(1)}$,..., $X_{U(n)}$.

REFERENCES

Aczel, S. (1966). Lectures on Functional Equations and Their Applications. Academic Press, New York.

Ahsanullah, M. (1977). A characteristic property of the exponential distribution. Ann. Statist. 5(3), 580-582.

Ahsanullah, M. (1978). Record Values and the Exponential Distribution. Ann. Inst. Stat. Math. 30, A, 429-433.

Ahsanullah, M. (1980). Linear prediction of record values for the two parameter exponential distribution. Ann. Inst. Statist. Math. 32, A, 363-368.

Ahsanullah, M. (1986), Estimation of parameters of a rectangular distribution by record values. Comp. Stat. Quarterly 2, 119-125.

Ahsanullah, M. (1995 a). Record Statistics. Nova Science Publishers, Inc. Commack, NY.

Ahsanullah, M. (1995 b). The generalized order statistics and a characteristic property of the exponential distribution. Pakistan Journal of Statistics. Vol. 11(3) pp 215-218.

Ahsanullah, M. (1996 a). The generalized order statistics from uniform distribution. Commun. Statist,-Theory Meth., 25(10),2311-2318.

Ahsanullah, M. (1996 b). Linear Prediction of generalized Order Statistics from Two Parameter Exponential Distribution. J. Appl. Statist. Sc, Vol. 3(1) 1-9.

Arnold, B.C. , Balakrishnan, N. and Nagaraja, H.N. (1992). A First Course in Order Statistics. John Wiley and Sons, Inc. New York.

Azlarov, T. A. and Volodin, N.A. (1986). Characterization problems Associated with the exponential distribution, Springer, New York.

David, H.A. (1981). Order Statistics (second edition). Wiley, New York.

Galambos, J. (1975). Characterizations of probability distributions by properties of order statistics I. In Statistical Distribution in Scientific Work. Ed. G.P. Patil et al. Vol. 3. D. Reidel Publishing Company, Dordrecht Holland, 71-88.

Gajek, L. and Gather, U. (1987). Characterizations of the exponential distribution by failure rate and moment properties of order statistics. Lecture Notes in Statistics # 51, 114-124. Springer- Verlag, New York.

Gradlsteyn, I.S. and Ryzhik, I.M. (1980). Tables of Integrals, Series and Products. Academic Press Inc. New York.

Graybill, F. A.(1969). Introduction to Matrices with Applications in Statistics . Wardsworth Publishing Company,Inc. Belmont, California.

Kamps, U. (1995). A Concept of Generalized Order Statistics. B. G. Teubner Stuttgart, Germany.

Applied Statistical Science, II
ISBN 1-56072-469-2

SOME INFERENCE ASPECTS OF A SOCIAL NETWORK

Bikas K. Sinha
Stat-Math Division
Indian Statistical Institute, Calcutta

1 INTRODUCTION

In a social network, we deal with binary square matrices $A = ((a_{ij}))$, $a_{ij} = 0$ or 1, $1 \leq i \neq j \leq N$ where a_{ij} and a_{ji} are not necessarily equal. The diagonal elements are not defined. Here, N represents the size of a finite population. The row totals

$$\{R_i = \sum_{\substack{j=1 \\ j(\neq i)}}^{N} a_{ij} \mid 1 \leq i \leq N\} \tag{1}$$

represent the out-degrees and the column totals

$$\{C_i = \sum_{\substack{i=1 \\ i(\neq j)}}^{N} a_{ij} \mid 1 \leq j \leq N\} \tag{2}$$

represent the in-degrees of the elements of the population. The grand total is given by

$$G = \sum R_i = \sum C_j \tag{3}$$

Clearly, the average out-degree and the average in-degree are equal

$$\overline{R} = \overline{C} = G/N \tag{4}$$

Technically, '$a_{ij} = 1$' represents a directed tie originating at i and terminating at $j : i \rightsquigarrow j$ while '$a_{ij} = a_{ji} = 1$' represents a symmetric tie : $i \leftrightsquigarrow j$. The average number of symmetric ties is defined as

$$\overline{I} = \frac{\sum\sum_{i<j} I(i,j)}{\binom{N}{2}}$$

where

$$\begin{aligned} I(i,j) &= 1 \quad \text{when } a_{ij}.\ a_{ji} = 1 \\ &= 0 \quad \text{otherwise} \end{aligned} \tag{5}$$

The problem of efficient estimation of $\overline{I}$ was addressed in Goswami et al. (1990). Here we propose to investigate the problem of unbiased estimation of $\overline{R}$ using various sampling strategies. The peculiar nature of the set-up calls for an understanding as to the type of data to be made available. This is discussed in Section 2. Detailed descriptions of some sampling strategies for sample size two are discussed in Section 3. An expository article in the area of social networks is Rao and Bandopadhyay (1987). Other related papers are also listed at the end.

2 TYPES OF DATA

We are permitted to sample any n elements of the population - either all at a time or sequentially, n being a specified integer. The ith element of the population, if sampled, may release :

$$\text{(i) the entire response vector} (a_{i1}, a_{i2}, \cdots, a_{ii-1}, a_{ii+1}, \cdots, a_{iN}) \tag{6}$$

$$\text{(ii) only the out-degree value } R_i = \sum_{j(\neq i)} a_{ij} \tag{7}$$

$$\text{(iii) a part of the response vector : } \{a_{ij} | j(\neq i) \quad \epsilon \quad \text{sample} \tag{8}$$

Clearly, (6) captures (7) while (8) provides the least. Several interesting questions arise :

Q1. Do we gain by utilizing (6) rather than (7) ?

Q2. Is (8) useful in any way for unbiased estimation of $\overline{R}$?

Q3. How does (8) compare with (6) and/or (7) ?

If sequential sampling is permitted, detailed information such as (6) released by the first member of the sample may be utilized in developing a technique for selecting the second member of the sample and so on. This will result in an ordered sample and Murthy's unordering principle will provide improved strategy (Murthy (1957)). The case of $n = 2$ highlights various interesting issues and is taken up in Section 3.

3 SAMPLING STRATEGIES FOR A PAIR OF UNITS

Strategy 1. Draw srswor sample of 2 units. Let $s = (i, j)$ be the sample. Observe (7) and obtain R_i and R_j. Propose

$$e_1(i, j) = \frac{R_i + R_j}{2}. \tag{9}$$

as an estimator for $\overline{R}$. Clearly,

$$E(e_1) = \overline{R}, \quad V_1 = V(e_1) = (\frac{1}{2} - \frac{1}{N})\frac{\sum_1^N (R_i - \overline{R})^2}{N-1} \tag{10}$$

Strategy 2. Draw srswor sample $s = (i, j)$. Observe (8) i.e. only the a_{ij} and a_{ji} values. Propose

$$e_2(i,j) = (N-1)\{\frac{a_{ij} + a_{ji}}{2}\} \tag{11}$$

as an estimator for $\overline{R}$.

Strategy 3. Assume $a_{ij}a_{ji} = a_{ij} = a_{ji} \quad \forall \; 1 \leq i < j \leq N$ (which holds for a communication network, for example).

Draw one unit at random. Call it i. Observe (6). Let $\mathcal{P}_i = \{j | a_{ij} = 1\}$ so that $|\mathcal{P}_i| = R_i$. Assume $R_i > 0 \; \forall i$. Select $j \in \mathcal{P}_i$ at random. Observe $\mathcal{P}_j$ and count R_j. Propose

$$e_3(i,j) = \frac{2R_iR_j}{R_i + R_j} = \text{H.M. of } R_i \text{ and } R_j \tag{12}$$

as an estimator of $\overline{R}$.

Strategy 4. Assume $a_{ij}a_{ji} = a_{ij} = a_{ji} \;\forall\; 1 \leq i < j \leq N$.

Draw unit i at random. Observe (6) and let $\overline{\mathcal{P}}_i = \{j | a_{ij} = 0\}$ so that $|\overline{\mathcal{P}}_i| = N-1-R_i$. Assume $R_i < N-1 \; \forall i$. Select, $j \in \overline{\mathcal{P}}_i$ at random. Observe $\mathcal{P}_j$ and count R_j. Propose

$$e_4(i,j) = \frac{(N-1)(R_i + R_j) - 2R_iR_j}{2(N-1) - R_i - R_j} \tag{13}$$

as an estimator of $\overline{R}$.

Strategy 1 is well-known. Strategy 2 uses the least data set. Strategies 3 and 4 are comparable to Strategy 1 since they both use R_i and R_j.

Below we compare these strategies.

Theorem 1.

$$E(e_2(i,j)) = \overline{R} \tag{14}$$

$$V_2 = V(e_2) = \frac{(N-1)\overline{R}}{2} + \frac{N-1}{2N}\sum\sum_{i \neq j} a_{ij}a_{ji} - \overline{R}^2. \tag{15}$$

Proof.

$$\begin{aligned} E(e_2) &= \frac{N-1}{2}\sum\sum_{i \neq j} \frac{a_{ij} + a_{ji}}{N(N-1)} \\ &= \frac{1}{2N}\sum\sum_{i \neq j}(a_{ij} + a_{ji}) = \frac{1}{N}\sum_i \sum_{j(\neq i)} a_{ij} = \overline{R}. \end{aligned}$$

$$\begin{aligned} E(e_2^2) &= \frac{(N-1)^2}{4N(N-1)} \sum\sum_{i \neq j} (a_{ij} + a_{ji})^2 \\ &= \frac{(N-1)^2}{4N(N-1)} \sum\sum_{i \neq j} (a_{ij} + a_{ji} + 2a_{ij}a_{ji}) \\ &= \frac{N-1}{2}\overline{R} + \frac{N-1}{2N} \sum\sum_{i \neq j} a_{ij}a_{ji}. \end{aligned}$$

which yields (15).

Theorem 2.

$$E(e_3(i,j)) = \overline{R} \tag{16}$$

$$V_3 = V(e_3) = \frac{2}{N} \sum\sum_{i<j} (a_{ij} \times a_{ji}) \text{H.M.}(R_i, R_j) - \overline{R}^2 \tag{17}$$

Proof. Note that the ordered strategy

$$\hat{\overline{R}}_3(i,j) = R_i, \quad p_3(i,j) = \frac{1}{NR_i} \tag{18}$$

is unbiased for $\overline{R}$.

The unordered version of $\hat{\overline{R}}_3(i,j)$ is given by (Vide Murthy (1957))

$$\tilde{\overline{R}}_3(i,j) = \frac{\frac{2}{N}}{\frac{1}{NR_i} + \frac{1}{NR_j}} = \frac{2R_iR_j}{R_i + R_j} \qquad \text{which coincides with (12).}$$

Note that $\tilde{p}_3(i,j) = \frac{R_i+R_j}{NR_iR_j}$ for every pair (i,j) for which $a_{ij} = a_{ji} = 1$.
Again,

$$\begin{aligned} E(e_3^2) &= \sum\sum_{i<j} \frac{4R_i^2R_j^2}{(R_i+R_j)^2} \times \frac{(R_i+R_j)}{NR_iR_j} a_{ij}a_{ji} \\ &= \frac{2}{N} \sum\sum_{i<j} (a_{ij} \times a_{ji}) H.M.(R_i, R_j) \end{aligned} \tag{19}$$

which yields (17) for V_3.

Theorem 3.

$$E(e_4(i,j)) = \overline{R} \tag{20}$$

$$V_4 = V(e_4) = \sum_{i<j}^{N}\sum \frac{\{(N-1)(R_i+R_j) - 2R_iR_j\}^2(1-a_{ij})(1-a_{ji})}{N(N-1-R_i)(N-1-R_j)\{2(N-1) - R_i - R_j\}} - \overline{R}^2 \tag{21}$$

Proof. Note that the ordered strategy

$$\hat{\overline{R}}_4(i,j) = R_i, \quad p_4(i,j) = \frac{1}{N(N-1-R_i)} \tag{22}$$

is unbiased for $\overline{R}$.

The unordered version of $\hat{\overline{R}}_4(i,j)$ is given by

$$\tilde{\overline{R}}_4(i,j) = \frac{\frac{R_i}{N-1-R_i} + \frac{R_j}{N-1-R_j}}{\frac{1}{N-1-R_i} + \frac{1}{N-1-R_j}} \qquad \text{which simplifies to } e_4 \text{ in (13).}$$

Note that

$$\tilde{p}_4(i,j) = \frac{2(N-1) - R_i - R_j}{N(N-1-R_i)(N-1-R_j)}$$

for every pair (i,j) for which $a_{ij} = a_{ji} = 0$.

Expression for V_4 follows from computation of $E(e_4^2)$.

Relative performances of the above strategies are examined below.

Theorem 4. (a) $V_1 \leq V_2$ whenever $a_{ij} \times a_{ji} = a_{ij} = a_{ji} \quad \forall i,j$
(b) $V_2 < V_1$ whenever there are only one-way ties between <u>any</u> two population units.

Proof.

(a) When $a_{ij}a_{ji} = a_{ij} = a_{ji}\ \forall i,j$, V_2 in (15) reduces to $(N-1)\overline{R} - \overline{R}^2$. In that case,

$$V_1 \leq V_2 \Leftrightarrow (N-2)\sum_1^N R_i^2 \leq N\overline{R}\{2(N-1)^2 - N\overline{R}\}$$

which follows by replacing $\sum_1^N R_i^2$ by its upper bound $N(N-1)\overline{R}$ and then by using the fact that $\overline{R} \leq N-1$. The stated condition implies that between any two members, either there is <u>no</u> tie at all or it is a two-way tie.

(b) If there are <u>no</u> reciprocities among the population units, then $a_{ij}a_{ji} = 0\ \forall i \neq j$. Hence, V_2 simplifies to $\frac{(N-1)\overline{R}}{2} - \overline{R}^2$ and in that case,

$$V_2 \leq V_1 \Leftrightarrow \frac{(N-1)\overline{R}}{2} \leq \frac{(N-2)\Sigma R_i^2}{2N(N-1)} + \frac{N\overline{R}^2}{2(N-1)}$$

$\Leftrightarrow \sigma^2 \geq \frac{\overline{R}(N-1)(N-1-2\overline{R})}{(N-2)}$ where σ^2 = variance of out-degrees.

Note that in view of $a_{ij}a_{ji} = 0$, we have $a_{ij} + a_{ji} \leq 1$ and hence $\overline{R} \leq (N-1)/2$. When $\overline{R} = (N-1)/2$, then above condition is <u>trivially</u> satisfied. Even when $\overline{R} < (N-1)/2$, σ^2 may be larger than the rhs quantity. The condition $a_{ij} + a_{ji} = 1$ corresponds to the fact that there is only one way tie between i and j. Below we give an illustrative example to the effect that $a_{ij} + a_{ji} \leq 1$ may lead to $V_2 < V_1$.

Example 1 $N = 8; a_{ij} = 1$ for

$$\begin{aligned}
&i = 1, j = 2, 3, 4, 5, 6, 8; \\
&i = 2, j = 3, 4, 5, 6; \\
&i = 3, j = 4, 5, 6, 7; \\
&i = 4, j = 6, 7, 8; \\
&i = 5, j = 4, 6, 7; \\
&i = 6, j = 7, 8; \\
&i = 7, j = 8; \\
&i = 8, j = 2, 3, 5.
\end{aligned}$$

For this example, $\overline{R} = 26/8 = 3.25; \sum R_i^2 = 100, \sigma^2 = 31/16, RHS = 91/48 < 31/16; a_{ij} + a_{ji} = 1$ except for (1,7) and (2,7) where it is zero.

Remark 1. In case of a communication network, either $a_{ij} = a_{ji} = 1$ or $a_{ij} = a_{ji} = 0$ and, hence, in such cases $V_1 < V_2$. In case of exactly one-way tie between pairs of elements (for example, in a hierarchical set-up), $V_2 < V_1$ and, hence, the least data set in (8) provides a better estimator ! If there is at the most one-way tie between pairs of elements, V_2 is likely to be smaller than V_1.

Theorem 5

(a) $V_3 \leq V_1$ whenever $\sum\sum_{i \neq j} a_{ij}a_{ji}\frac{(R_i - R_j)^2}{R_i + R_j} \geq \sum\sum_{i \neq j}(R_i - R_j)^2/2(N-1)$.

(b) For a wheel-shaped communication network, $V_3 < V_1$ for $N > 4$.

(c) For a communication network with only two out-degree values R(freq.1) and r(freq. $N-1$), $V_3 < V_1$ whenever $R > r$.

Proof

(a) Referring to (17) and (10),

$$\begin{aligned}
V_3 \leq V_1 \iff &\frac{2}{N}\sum\sum_{i<j} a_{ij}a_{ji}H.M.(R_i, R_j) \leq \frac{(N-2)\sum R_i^2}{2N(N-1)} + \frac{N\overline{R}^2}{2(N-1)} \\
\iff &\frac{2}{N}\sum\sum_{i<j} a_{ij}a_{ji}A.M.(R_i, R_j) \leq \frac{(N-2)\sum R_i^2}{2N(N-1)} + \frac{N\overline{R}^2}{2(N-1)}
\end{aligned}$$

$$+\frac{2}{N}\sum\sum_{i<j} a_{ij}a_{ji}[AM(R_i, R_j) - H.M.(R_i, R_j)]$$

$$\Longleftrightarrow \quad \frac{1}{N}\sum\sum_{i\neq j} a_{ij}a_{ji}(\frac{R_i + R_j}{2}) \leq \cdots$$

$$\Longleftrightarrow \quad \frac{1}{N}\sum R_i^2 \leq \cdots$$

$$\Longleftrightarrow \quad \frac{1}{N}\sum\sum_{i\neq j} a_{ij}a_{ji}[A.M.(R_i, R_j) - H.M.(R_i, R_j)]$$

$$\geq \frac{1}{N}\sum R_i^2 - \frac{N-2}{2N(N-1)}\sum R_i^2 - \frac{N\overline{R}^2}{2(N-1)} = \frac{N\sigma^2}{2(N-1)} \qquad \text{(upon simplification)}$$

$$\Longleftrightarrow \frac{1}{N}\sum\sum_{i\neq j} a_{ij}a_{ji}\frac{(R_i - R_j)^2}{2(R_i + R_j)} \geq \frac{\sum\sum_{i\neq j}(R_i - R_j)^2}{4N(N-1)}$$

(using the representation $\sigma^2 = \sum\sum_{i\neq j}(R_i - R_j)^2/2N^2$)

$$\Longleftrightarrow \sum\sum_{i\neq j} a_{ij}a_{ji}\frac{(R_i - R_j)^2}{R_i + R_j} \geq \frac{\sum\sum_{i\neq j}(R_i - R_j)^2}{2(N-1)} \qquad (23)$$

which proves (a).

(b) Next, consider a wheel-shaped network with the central point having serial number N. Then $R_1 = \cdots = R_{N-1} = 3$ and $R_N = N - 1$. Then, in (23),

$$\text{LHS} = \frac{(N-4)^2(N-1)}{N+2} \quad \text{and}$$

$$\text{RHS} = \frac{(N-1)(N-4)^2}{2(N-1)} = \frac{(N-4)^2}{2} \qquad \text{whence (b) follows.}$$

(c) Under the stated conditions, LHS $= \frac{R(R-r)^2}{R+r}$ while RHS $= \frac{(R-r)^2(N-1)}{2(N-1)} = \frac{(R-r)^2}{2}$. Hence, $V_3 < V_1$ whenever $R > r$.

An example of a network of the type described in (c) is given below.
Example 2.
$N = 7$ and $a_{ij} = a_{ji} = 1$ for

$$i = 1, j = 2, 3, 4, 5, 6, 7;$$
$$i = 2, j = 3, 4;$$
$$i = 3, j = 1, 2;$$
$$i = 4, j = 2, 5;$$

$$+\frac{2}{N}\sum\sum_{i<j} a_{ij}a_{ji}[AM(R_i,R_j) - H.M.(R_i,R_j)]$$

$$\Longleftrightarrow \quad \frac{1}{N}\sum\sum_{i\neq j} a_{ij}a_{ji}(\frac{R_i+R_j}{2}) \leq \cdots$$

$$\Longleftrightarrow \quad \frac{1}{N}\sum R_i^2 \leq \cdots$$

$$\Longleftrightarrow \quad \frac{1}{N}\sum\sum_{i\neq j} a_{ij}a_{ji}[A.M.(R_i,R_j) - H.M.(R_i,R_j)]$$

$$\geq \frac{1}{N}\sum R_i^2 - \frac{N-2}{2N(N-1)}\sum R_i^2 - \frac{N\overline{R}^2}{2(N-1)} = \frac{N\sigma^2}{2(N-1)} \qquad \text{(upon simplification)}$$

$$\Longleftrightarrow \frac{1}{N}\sum\sum_{i\neq j} a_{ij}a_{ji}\frac{(R_i-R_j)^2}{2(R_i+R_j)} \geq \frac{\sum\sum_{i\neq j}(R_i-R_j)^2}{4N(N-1)}$$

(using the representation $\sigma^2 = \sum\sum_{i\neq j}(R_i - R_j)^2/2N^2$)

$$\Longleftrightarrow \sum\sum_{i\neq j} a_{ij}a_{ji}\frac{(R_i-R_j)^2}{R_i+R_j} \geq \frac{\sum\sum_{i\neq j}(R_i-R_j)^2}{2(N-1)} \qquad (23)$$

which proves (a).

(b) Next, consider a wheel-shaped network with the central point having serial number N. Then $R_1 = \cdots = R_{N-1} = 3$ and $R_N = N-1$. Then, in (23),

$$\text{LHS} = \frac{(N-4)^2(N-1)}{N+2} \quad \text{and}$$

$$\text{RHS} = \frac{(N-1)(N-4)^2}{2(N-1)} = \frac{(N-4)^2}{2} \qquad \text{whence (b) follows.}$$

(c) Under the stated conditions, LHS $= \frac{R(R-r)^2}{R+r}$ while RHS $= \frac{(R-r)^2(N-1)}{2(N-1)} = \frac{(R-r)^2}{2}$. Hence, $V_3 < V_1$ whenever $R > r$.

An example of a network of the type described in (c) is given below.
Example 2.
$N = 7$ and $a_{ij} = a_{ji} = 1$ for

$i = 1, j = 2, 3, 4, 5, 6, 7;$
$i = 2, j = 3, 4;$
$i = 3, j = 1, 2;$
$i = 4, j = 2, 5;$

$$i = 5, j = 1, 4;$$
$$i = 6, j = 1, 7;$$
$$i = 7, j = 1, 6.$$

In this example, $R = 4$ and $r = 2$ and $V_3 < V_1$.

Theorem 6. $e_4 = \overline{R}\ \forall$ pairs (i, j) whenever the population splits into two clusters C_1 and C_2 with $|C_1| = N_i$, $i = 1, 2$; $j \in C_i \Rightarrow R_j = N_i - 1$ $\forall\, j \in C_i$, $i = 1, 2$.

Proof. Note that $e_4(i, j)$ is defined for $i \in C_1$ and $j \in C_2$ or for $i \in C_2$ and $j \in C_1$ and is the same for both. Referring to (13), we have

$$\begin{aligned} e_4(i,j) &= \frac{(N-1)(N_1 - 1 + N_2 - 1) - 2(N_1 - 1)(N_2 - 1)}{2(N-1) - (N_1 - 1) - (N_2 - 1)} \\ &= \frac{N^2 - N - 2N_1N_2}{N} = \frac{N_1^2 + N_2^2 - N}{N} \end{aligned}$$

Again, $\overline{R} = \frac{N_1(N_1-1)+N_2(N_2-1)}{N} = \frac{N_1^2+N_2^2-N}{N}$ and this coincides with e_4.

Remark 2. It is possible to construct examples where the population splits into 3 clusters and yet e_4 is better than e_1 ! Here is an example.
Example 3.
$N = 9$ and $a_{ij} = a_{ji} = 1$ for

$$i = 1, j = 2;$$
$$i, j) = 3, 4, 5$$
$$and(i, j) = 6, 7, 8, 9.$$

s	$p_4(s)$	$e_4(s)$
$C_1 \cap C_2$	13/63	20/13
$C_1 \cap C_3$	32/105	13/6
$C_2 \cap C_3$	22/45	28/11

$$N = 9,\ \overline{R} = 20/9,\ V_1 = \frac{175}{648} \doteq 0.2700$$

$$E(e_4) = 20/9,\ V_4 \doteq 0.1486 << V_1!!$$

References

Achuthan, N., Rao, S.B. and Rao, A.R. (1982). The number of symmetric edges in a digraph with prescribed out-degrees. In *CombinatoricsandApplications*, Proc. of a Seminar held in honour of Prof. S. S. Shrikhande, Eds: K.S. Vijayan and N.M. Singhi, ISI, Calcutta.

Bandyopadhyay, S., Rao, A.R. and Sinha, B.K. (1993). Methodology for studying Social Transformation: Use of Social Networks. Presented in the Symposium on *SocialTransformation : DifferentDimensions*, as a part of P.C. Mahalanobis Centenary Celebrations, held at Giridih.

Chatterjee, A.K., Bandyopadhyay, S. and Rao, A.R. (1993). Relative importance of different factors for boundary of reciprocity: An illustration. *Connections*, 16, 15-22.

Goswami, A., Sinha, B.K. and Sengupta, S. (1990). Optimal strategies in sampling from a social network. *SequentialAnalysis*, 9, 1-18.

Murthy, M.N. (1957). Order and unordered estimators in sampling without replacement. *Sankhya*, 18, 379-390.

Rao, A.R. and Bandyopadhyay, S. (1987). Measures of reciprocity in a social network. *Sankhya*, Ser. A, 49, 141-188.

Rao, A.R., Jana, R. and Bandyopadhyay, S. (1996). A Markov Chain Monte Carlo method for generating random (0,1)-matrices with given marginals. *Sankhya*, Ser. A, 58, 225-242.

Applied Statistical Science, II
ISBN 1-56072-469-2

ESTIMATION OF MEAN BASED ON AN UNBALANCED RANKED SET SAMPLE

Philip L.H. Yu
Kin Lam
Department of Statistics
University of Hong Kong

Bimal K. Sinha
Department of Math/Statistics
University of Maryland Baltimore County
Baltimore, Maryland
USA

SUMMARY

In this paper we address the problem of estimation of the mean of a population based on an *unbalanced* ranked set sample (RSS), which is a modification of the original RSS of McIntyre (1952).

KEY WORDS : Normal distribution; Ranked set sample; Simple random sample; *t*-distribution; Unequal allocation.

1. INTRODUCTION

The basic concept behind a ranked set sample (*RSS*) can be briefly described as follows. Suppose $X_1, X_2, \cdots, X_n$ is a random sample from $F(x)$ with a mean μ and a finite variance σ^2. Then a standard nonparametric estimator of μ is $\bar{X} = \sum_1^n X_i/n$ with $var(\bar{X}) = \sigma^2/n$. In contrast to *SRS*, *RSS* uses only one observation, namely, $X_{1:n} \equiv X_{(11)}$, the lowest observation, from this set, then $X_{2:n} \equiv X_{(22)}$, the second lowest from another independent set of n observations, and finally $X_{n:n} \equiv X_{(nn)}$, the largest observation from a last set of n observations. This process can be described in a table as follows.

Table 1.1. Display of n^2 observations in n sets of n each

$X_{(11)}$	$X_{(12)}$	$\cdots$	$X_{(1(n-1))}$	$X_{(1n)}$
$X_{(21)}$	$X_{(22)}$	$\cdots$	$X_{(2(n-1))}$	$X_{(2n)}$
$\vdots$	$\vdots$		$\vdots$	$\vdots$
$X_{(n1)}$	$X_{(n2)}$	$\cdots$	$X_{(n(n-1))}$	$X_{(nn)}$

The important point to emphasize is that although *RSS* requires identification of as many as n^2 experimental or sampling units, only n of them, namely, $\{X_{(11)}, \cdots, X_{(nn)}\}$, are actually measured, thus making a comparison of this sampling strategy with *SRS* of the same size n meaningful. Obviously, *RSS* would be a serious contender to *SRS* in situations where the task of assembly of the sampling units is easy and their relative rankings in terms of the characteristic under study can be done with negligible cost. The new sample $X_{(11)}, X_{(22)}, ..., X_{(nn)}$, known in the literature as a *Ranked Set Sample* (*RSS*), are independent but not identically distributed. Moreover, marginally, $X_{(ii)}$ is distributed as $X_{i:n}$, the ith order statistic in a sample of size n from $F(x)$.

McIntyre (1952) proposed

$$\hat{\mu}_{rss} = \sum_1^n X_{(ii)}/n \tag{1.1}$$

as a rival estimator of μ as opposed to $\bar{X}$. It is easy to verify that $E(\hat{\mu}_{rss}) = \mu$, and a somewhat surprising result (Takahasi and Wakamoto, 1968) which makes *RSS* a serious contender is that

$$var(\hat{\mu}_{rss}) < var(\bar{X}) \ ! \tag{1.2}$$

Dell (1969) and Dell and Clutter (1972) provided the following explicit expression for the variance of $\hat{\mu}_{rss}$ where $\mu_{(i)}$ is the mean of $X_{i:n}$.

$$var(\hat{\mu}_{rss}) = \sigma^2/n - \sum_1^n(\mu_{(i)} - \mu)^2/n^2. \tag{1.3}$$

Many aspects of RSS have been studied in the literature, and we refer to Sinha et al. (1996) for references.

Since an application of RSS involves assembly of n^2 units and repeated ranking of n units at a time, naturally it is most suitable for small values of n. However, this has the negative implication that the resultant RSS estimator may not be very efficient. Moreover, the SRS-based estimator, $\bar{X}$, does not suffer from such artificial constraints. To increase the efficiency of the RSS-based estimator of μ, McIntyre (1952) suggested replicating the entire RSS process several times. Thus, for example, to compare against an SRS of size 15, instead of working with an RSS based on $n = 15$ which requires assembly of 225 units and ranking 15 units at a time, which may involve ranking errors, one may consider an RSS procedure based on 3 units and replicating the whole process 5 times, thus requiring an assembly of 45 units and a total of 15 actual measurements as under the SRS scheme. Another possibility is to use an RSS procedure based on 5 units and replicating the entire process 3 times. Quite generally, if $n = r \times s$, with $r \leq s$, we can use an RSS procedure based on r units at a time, and repeat the process s times. This would require an assembly of $r^2 s$ units and actual measurements of n units as under the SRS scheme. Referring to Table 1.1 with $n = r$, the data collected from the ith cycle can be denoted by $\{X_{(11:r)}^{(i)}, \cdots, X_{(rr:r)}^{(i)}\}$, $i = 1, \cdots, s$. The overall estimator of μ is then given by

$$\begin{aligned} \hat{\mu}_{rss}(n = r \times s) &= \sum_{i=1}^{i=s}[\sum_{j=1}^{j=r} X_{(jj:r)}^{(i)}/r]/s \\ &= \sum_{j=1}^{j=r}[\sum_{i=1}^{i=s} X_{(jj:r)}^{(i)}/s]/r. \end{aligned} \tag{1.4}$$

The expression within [.] in the first part of (1.4) is clearly the average of r order statistics for a fixed cycle, while that in the last part of (1.4) is the average of the jth order statistic (in a sample of size r) over s cycles. The variance of $\hat{\mu}_{rss}(n = r \times s)$ is given by (see (1.3))

$$var(\hat{\mu}_{rss}(n = r \times s)) = [\sigma^2/r - \sum_{j=1}^{j=r}(\mu_{(j:r)} - \mu)^2/r^2]/s, \tag{1.5}$$

where $E(X_{(jj:r)}) = \mu_{(j:r)}$, thus indicating its superiority over $\bar{X}$ based on $n = r \times s$ units. Writing

$$var(X_{(jj:r)}) = \sigma^2_{(j:r)}, \tag{1.6}$$

it follows easily from (1.4) that

$$E(\hat{\mu}_{rss}(n = r \times s)) = \sum_{j=1}^{j=r} \mu_{(j:r)}/r = \mu, \quad var(\hat{\mu}_{rss}(n = r \times s)) = \sum_{j=1}^{j=r} \sigma^2_{(j:r)}/r^2 s. \tag{1.7}$$

Comparing (1.5) and (1.7), we immediately obtain

$$\frac{\sum_{j=1}^{j=r} \sigma^2_{(j:r)}}{r} \leq \sigma^2. \tag{1.8}$$

What we have described above can be called an *equal* allocation scheme in the sense that each of the r order statistics is replicated an equal number of times, namely, s times. It is quite possible to use *unequal* allocation schemes as well. Starting with r units at a time and measuring (after ranking) the smallest unit s_1 times, the second smallest s_2 times, and so on, we end up with a collection of $n = s_1 + \cdots + s_r$ measurements. An unbiased estimator of μ is then constructed as (see the second expression in (1.4))

$$\hat{\mu}_{rss}(unequal) = \sum_{j=1}^{j=r} [\sum_{i=1}^{s_j} X^{(i)}_{(jj:r)}/s_j]/r \tag{1.9}$$

with its variance given by

$$var(\hat{\mu}_{rss}(unequal)) = \frac{1}{r^2} \sum_{j=1}^{j=r} \sigma^2_{(j:r)}/s_j. \tag{1.10}$$

Clearly (1.10) reduces to (1.7) when $s_1 = \cdots = s_r = s$. One may ask if, for a fixed n, $\hat{\mu}_{rss}(unequal)$ is always better than $\bar{X}$, i.e., if

$$var(\hat{\mu}_{rss}(unequal)) \leq \sigma^2/n, \quad n = s_1 + \cdots + s_r. \tag{1.11}$$

It is easy to construct a counter-example which shows that this need not be the case. Taking $r = 2$ and writing $var(X_{(11)}) = \sigma^2_{(1:2)}$, $var(X_{(12)}) = \sigma^2_{(2:2)}$ and $cov(X_{(11)}, X_{(12)}) = \sigma_{(12:2)}$, it follows that the inequality in (1.11) above boils down to

$$\frac{s_2}{s_1}\sigma^2_{(1:2)} + \frac{s_1}{s_2}\sigma^2_{(2:2)} \leq \sigma^2_{(1:2)} + \sigma^2_{(2:2)} + 4\sigma_{(12:2)}. \tag{1.12}$$

In the above we have used the fact that $\sigma^2 = \frac{\sigma^2_{(1:2)} + \sigma^2_{(2:2)}}{2} + \sigma_{(12:2)}$. Clearly, (1.12) cannot hold for arbitrary s_1 and s_2. For example, when $\sigma_{(1:2)} = \sigma_{(2:2)}$ (this holds for a normal distribution), using the fact that $\sigma_{(12:2)} \leq \sigma^2_{(1:2)}$, a necessary condition for (1.12) to hold is that $\frac{s_2}{s_1} + \frac{s_1}{s_2} \leq 6$, which obviously may not be true.

On the other hand, it is possible to discuss the problem of *optimal* allocation of the replications $s_1, \cdots, s_r$ among the r order statistics for a fixed number n of actual measurements. It is readily seen from (1.10) that the optimal allocation corresponds to what is popularly known as *Neyman* allocation in the context of stratified sampling, and is given by (see also Patil et al. (1994))

$$s_j(optimal) = \frac{\sigma_{(j:r)}}{\sum_{j=1}^{j=r} \sigma_{(j:r)}} n, \tag{1.13}$$

with the resultant optimal variance as

$$var(\hat{\mu}_{rss}(unequal/optimal)) = \frac{(\sum_{j=1}^{j=r} \sigma_{(j:r)})^2}{nr^2}. \tag{1.14}$$

A comparison of (1.8) and (1.14) immediately reveals that this optimal variance is smaller than $var(\bar{X})$, as expected. We therefore conclude that although an RSS with an unequal allocation in general need not be more efficient than an SRS with the same number of actual measurements, whenever the optimal allocation displayed in (1.13) can be implemented, it will result in a better strategy. It is not difficult to verify that any parent distribution $F(x)$ belonging to a *known* location and scale family with a finite variance would render the optimal allocation completely known, and hence useful. This is primarily because $\sigma^2_{(j:r)} = \sigma^2 \nu_{jj:r}$ where $\nu_{jj:r}$ is the variance of the jth order statistic in a sample of size r from the original parent distribution with location 0 and scale unity.

It should be noted however that when the parent distribution $F(x)$ belongs to a known location and scale family with a finite variance, it is indeed possible to use a better estimator of the mean μ as opposed to $\hat{\mu}_{rss}(unequal)$ given in (1.9). Writing $\bar{X}_{(jj:r)} = \sum_{i=1}^{i=s_j} X^{(i)}_{(jj:r)}/s_j$, and noting that

$$E(\bar{X}_{(jj:r)}) = \mu + \sigma\nu_{j:r}, \quad var(\bar{X}_{(jj:r)}) = \sigma^2\nu_{jj:r}/s_j, \quad j = 1, \cdots, r, \tag{1.15}$$

it follows immediately that the $BLUE$ of μ based on $\bar{X}_{(jj:r)}$'s is given by

$$\hat{\mu}_{blue}(unequal) = \lambda_1 \sum_{j=1}^{j=r} \frac{s_j \bar{X}_{(jj:r)}}{\nu_{jj:r}} + \lambda_2 \sum_{j=1}^{j=r} \frac{s_j \nu_{j:r} \bar{X}_{(jj:r)}}{\nu_{jj:r}} \tag{1.16}$$

where

$$\lambda_1 = \frac{\sum_{j=1}^{j=r} \frac{s_j \nu^2_{j:r}}{\nu_{jj:r}}}{(\sum_{j=1}^{j=r} \frac{s_j}{\nu_{jj:r}})(\sum_{j=1}^{j=r} \frac{s_j \nu^2_{j:r}}{\nu_{jj:r}}) - (\sum_{j=1}^{j=r} \frac{s_j \nu_{j:r}}{\nu_{jj:r}})^2} \tag{1.17}$$

and

$$\lambda_2 = -\frac{\sum_{j=1}^{j=r} \frac{s_j \nu_{j:r}}{\nu_{jj:r}}}{(\sum_{j=1}^{j=r} \frac{s_j}{\nu_{jj:r}})(\sum_{j=1}^{j=r} \frac{s_j \nu_{j:r}^2}{\nu_{jj:r}}) - (\sum_{j=1}^{j=r} \frac{s_j \nu_{j:r}}{\nu_{jj:r}})^2}. \tag{1.18}$$

This results in the variance of $\hat{\mu}_{blue}(unequal)$ as

$$var(\hat{\mu}_{blue}(unequal)) = \frac{\sigma^2 \sum_{j=1}^{j=r} \frac{s_j \nu_{j:r}^2}{\nu_{jj:r}}}{(\sum_{j=1}^{j=r} \frac{s_j}{\nu_{jj:r}})(\sum_{j=1}^{j=r} \frac{s_j \nu_{j:r}^2}{\nu_{jj:r}}) - (\sum_{j=1}^{j=r} \frac{s_j \nu_{j:r}}{\nu_{jj:r}})^2}. \tag{1.19}$$

It is again interesting to verify if $\hat{\mu}_{blue}(unequal)$, defined in (1.16), is better than $\bar{X}$ for *all* allocations $s_1, \cdots, s_r$ such that $s_1 + \cdots + s_r = n$. Assume a symmetric allocation: $s_j = s_{r-j+1}$. Then the results we obtained here are the following.

Result 1. When $r = 2$, it is always true, i.e., $var(\hat{\mu}_{blue}(unequal)) \leq \sigma^2/n$ for any symmetric allocation of s_j and any underlying distribution.

Proof. Clearly when $r = 2$, symmetric allocation of s_j implies that $var(\hat{\mu}_{blue}(unequal)) = var(\hat{\mu}_{blue}(equal)) \leq var(\bar{X}) = \sigma^2/n$.

Result 2. For a normal distribution, it is always true for any r and any symmetric allocation of s_j.

Proof. From (1.19), we find that for a normal distribution, whenever a symmetric allocation is used,

$$var(\hat{\mu}_{blue}(unequal)) = \sigma^2/(\sum_{j=1}^{j=r} s_j/\nu_{jj:r}). \tag{1.20}$$

It is easy to show that $var(\hat{\mu}_{blue}(unequal))$ is maximized when $s_1 = s_r = n/2$. Thus,

$$var(\hat{\mu}_{blue}(unequal)) \leq \nu_{11:r}\sigma^2/n. \tag{1.21}$$

Using Theorem 4.9.1 in Arnold et al.(1992) that $\sum_{j=1}^{j=r} \nu_{1j:r} = 1$, and the Fact 6.2.1 in Tong(1990) that the order statistics have non-negative correlations, it follows readily that $\nu_{11:r} \leq 1$ for any r. Hence the proof.

Result 3. We found numerically that *Result 2* is also true for a logistic and a uniform distribution, and we suspect that this is *always* true for any symmetric distribution.

Returning to (1.19), it is clear that we can easily carry out the computation of *optimum* allocations $s_1, \cdots, s_r$ for any *known* location and scale family of distribution with a finite variance. In Section 2 we specialize to the case of a normal distribution and use the previous results to estimate its mean. It turns out however that, for (1.9), unless n is very large the optimum replications $s_1, \cdots, s_r$ are nearly equal ! On the other hand, for (1.16), it turns out that the optimum allocation corresponds to as large an s_j as possible for the middle order statistics and as small an s_j as possible for other order statistics (see Sinha et al. (1996)). In Section 3 we carry out our analysis for a t-family of distributions with a known $d.f.$ ν, and demonstrate the role of ν in utilizing the unequal optimal allocations in the context of (1.9).

2. ESTIMATION OF A NORMAL MEAN

In this section we consider the problem of estimation of a normal mean based on optimal unequal replications. We have used Table 4.9.2 in Arnold et al. (1992) for values of $\nu_{jj:r}$ for various combinations of r and j.

Referring to equation (1.14), it is clear that we need to compare

$$Q_r = [\frac{\sum_{j=1}^{j=r} \sqrt{\nu}_{jj:r}}{r}]^2 \tag{2.1}$$

against 1 irrespective of the values of σ^2 and the sample size n. We have considered the cases $r = 2(1)8$. Table 2.1 presents the values of Q_r and also the values for equal replications for these values of r. It is easily seen that the variances under equal replications are almost as good as those under optimal unequal replications. Moreover, subject to the integer-valued restriction, the unequal replications essentially boil down to equal replications unless n is very large.

Table 2.1. The values of Q_r and that for equal replications

r	Q_r	$\sum_{j=1}^{j=r} \nu_{jj:r}/r$ (values for equal replications)
2	0.6817	0.6817
3	0.5212	0.5225
4	0.4235	0.4261
5	0.3576	0.3610
6	0.3098	0.3139
7	0.2736	0.2782
8	0.2452	0.2501

On the other hand, for (1.19), since the variance of the sample median is the smallest among the variances of all order statistics for a normal distribution (Sinha et al. (1996)), our computations for different values of r show that the optimal allocation corresponds to maximum s_j for the median and as small s_j as possible for other order statistics. However, since identifying the extremes is much easier compared to the middle order statistics, in Table 2.2 we have considered for $r = 3, 4, 5, 6$ and $n = 12$, various combinations of $s_1, \cdots, s_r$, and observed that even taking high values of s_j for the extremes and low values of s_j for the intermediate order statistics results in the dominance of the proposed estimator $\hat{\mu}_{blue}(unequal)$. It may be noted that in the case of a normal distribution, due to its symmetry, $\sum_{j=1}^{j=r} \frac{s_j \nu_{j:r}}{\nu_{jj:r}} = 0$, whenever the symmetry of the s_j's is invoked. The last entry in Table 2.2 represents the relative precision (RP) given by $var(\hat{\mu}_{blue}(unequal))/var(\bar{X}) = n/\sum_{j=1}^{j=r} \frac{s_j}{\nu_{jj:r}}$.

Table 2.2. The values of RP for various allocations $r = 3, 4, 5, 6$ and $n = 12$

r	s_1	s_2	s_3	s_4	s_5	RP	r	s_1	s_2	s_3	s_4	s_5	s_6	RP
3	0	12	0			0.4487	4	0	6	6	0			0.3605
	1	10	1			0.4640		1	5	5	1			0.3772
	2	8	2			0.4804		2	4	4	2			0.3957
	3	6	3			0.4980		3	3	3	3			0.4160
	4	4	4			0.5169		4	2	2	4			0.4385
	5	2	5			0.5374		5	1	1	5			0.4636
	6	0	6			0.5595		6	0	0	6			0.4917
5	0	0	12	0	0	0.2868	6	0	0	6	6	0	0	0.2462
	1	0	10	0	1	0.3051		1	0	5	5	0	1	0.2642
	1	1	8	1	1	0.3094		1	1	4	4	1	1	0.2699
	2	0	8	0	2	0.3258		2	0	4	4	0	2	0.2850
	2	1	6	1	2	0.3308		2	1	3	3	1	2	0.2917
	3	0	6	0	3	0.3496		3	0	3	3	0	3	0.3093
	2	2	4	2	2	0.3359		2	2	2	2	2	2	0.2987
	3	1	4	1	3	0.3553		3	1	2	2	1	3	0.3172
	4	0	4	0	4	0.3771		4	0	2	2	0	4	0.3382
	3	2	2	2	3	0.3612		3	2	1	1	2	3	0.3256
	4	1	2	1	4	0.3838		4	1	1	1	1	4	0.3477
	5	0	2	0	5	0.4093		5	0	1	1	0	5	0.3731
	3	3	0	3	3	0.3673		3	3	0	0	3	3	0.3344
	4	2	0	2	4	0.3907		4	2	0	0	2	4	0.3578
	5	1	0	1	5	0.4172		5	1	0	0	1	5	0.3847
	6	0	0	0	6	0.4475		6	0	0	0	0	6	0.4159

3. ESTIMATION OF MEAN FOR t: UNEQUAL REPLICATIONS

In this section we consider the situation when the underlying population is a member of the t-family, and explore if the use of an unequal RSS for estimation of its mean μ really makes a difference as compared to an equal allocation or balanced RSS. For $\nu \geq 3$, we take the *pdf* as

$$f(x|\mu,\eta) = \frac{1}{\sqrt{(\nu)}\eta B(1/2,\nu/2)}.[1+\frac{(x-\mu)^2}{\nu\eta^2}]^{-(\frac{\nu+1}{2})}, \quad -\infty < x, \mu < \infty, \quad \eta > 0 \tag{3.1}$$

so that $E(X) = \mu$ and $var(X) = \frac{\nu\eta^2}{\nu-2} = \sigma^2$. We assume all throughout that ν is known. Writing $X = \mu + \eta Y$, it follows readily that the optimal allocations $s_j(optimal)$'s given in (1.13) are completely known, and can be obtained from the known distribution of Y. Although not optimal, we have taken $\bar{X}$ as the estimator of μ under an SRS.

It follows from (1.13) that

$$s_j(optimal) = \frac{\sigma^*_{(j:r)}(\nu)}{\sum_{j=1}^{j=r}\sigma^*_{(j:r)}(\nu)}n \tag{3.2}$$

where $\sigma^*_{(j:r)}(\nu) = \sqrt{var(Y_{(j:r)})}$, and from (1.14) that

$$var(\hat{\mu}_{rss}(unequal/optimal)) = \frac{\eta^2(\sum_{j=1}^{j=r}\sigma^*_{(j:r)}(\nu))^2}{nr^2} \tag{3.3}$$

$$= \frac{\sigma^2}{n}.\frac{\nu-2}{\nu}.[\frac{\sum_{j=1}^{j=r}\sigma^*_{(j:r)}(\nu)}{r}]^2. \tag{3.4}$$

Keeping aside the common factor σ^2/n, we have evaluated the quantity

$$Q(r,\nu) = \frac{\nu-2}{\nu}.[\frac{\sum_{j=1}^{j=r}\sigma^*_{(j:r)}(\nu)}{r}]^2$$

for various values of r and ν in Table 3.1, based on the Tables in Tiku and Kumra (1985), and compared it against 1. We have also included the quantity $Q^*(r,\nu) = \frac{\nu-2}{\nu}\sum_{j=1}^{j=r}\nu_{jj:r}/r$ which obtain for equal allocations (balanced case). Evidently, a very large value of ν corresponds to a normal distribution. It is clear from Table 3.1 that the optimal unequal replications do play a role for small values of ν.

Table 3.1. The values of $Q(r,\nu)$ and $Q^*(r,\nu)$

r	ν	$Q(r,\nu)$	$Q^*(r,\nu)$	r	ν	$Q(r,\nu)$	$Q^*(r,\nu)$
3	3	0.6148	0.6581	4	3	0.5118	0.5865
	4	0.5740	0.5934		4	0.4753	0.5095
	5	0.5570	0.5690		5	0.4593	0.4807
	6	0.5481	0.5567		6	0.4507	0.4662
	7	0.5426	0.5495		7	0.4453	0.4576
	8	0.5390	0.5447		8	0.4417	0.4520
	9	0.5364	0.5413		9	0.4391	0.4480
	10	0.5344	0.5388		10	0.4371	0.4451
	19	0.5273	0.5299		19	0.4299	0.4347
	∞	0.5212	0.5225		∞	0.4235	0.4261
5	3	0.4396	0.5363	6	3	0.3861	0.4984
	4	0.4068	0.4515		4	0.3564	0.4084
	5	0.3920	0.4200		5	0.3426	0.3753
	6	0.3838	0.4042		6	0.3350	0.3588
	7	0.3787	0.3949		7	0.3301	0.3491
	8	0.3753	0.3888		8	0.3269	0.3427
	9	0.3728	0.3846		9	0.3245	0.3383
	10	0.3709	0.3814		10	0.3227	0.3350
	19	0.3638	0.3702		19	0.3159	0.3234
	∞	0.3576	0.3610		∞	0.3098	0.3139

ACKNOWLEDGEMENTS

The research of K. Lam was partially supported by Hong Kong RGC grant 334/017/0001.

REFERENCES

Arnold, B.C., Balakrishnan N. and Nagaraja, H.N. (1992). **A First Course in Order Statistics**. John Wiley.

David, H.A. and Levine, D.N. (1972), "Ranked set sampling in the presence of judgement error," *Biometrics*, 28, 553-555.

Dell, T.R. (1969), "*The theory of some applications of ranked set sampling,*" Ph.D. thesis, University of Georgia, Athens, GA.

Dell, T.R. and Clutter, J.L. (1972), "Ranked set sampling theory with order statistics background," *Biometrics*, 28, 545-555.

McIntyre, G.A. (1952), "A method for unbiased selective sampling, using ranked sets," *Australian J. Agricultural Research*, 3, 385-390.

Patil, G.P., Sinha, A.K. and Taillie, C. (1994). "Ranked Set Sampling" in **Handbook of Statistics, Volume 12**, 167-200. Edited by Patil and Rao, Elsevier Science B.V.

Sinha, Bimal K., Sinha, Bikas K. and Purkayastha, S. (1994), "On Some Aspects of Ranked Set Sampling for Estimation of Normal and Exponential Parameters", *Statistics & Decisions*, 14, 223-240.

Takahasi, K., and Wakimoto, K. (1968), "On unbiased estimates of the population mean based on the sample stratified by means of ordering," *Ann. Inst. Statist. Math.*, 20, 1-31.

Tiku, M.L. and Kumra, S. (1985). "Expected values and variances and covariances of order statistics for a family of symmetric distributions (student's t)". Selected Tables in Mathematical Statistics, Volume 8. Edited by the Institute of Mathematical Statistics. American Mathematical Society, Providence, Rhode Island.

Tong, Y.L. (1990). **The Multivariate Normal Distribution**. Springer-Verlag.

Applied Statistical Science, II
ISBN 1-56072-469-2

LOWER BOUND TO MEAN SQUARE ERROR OF ESTIMATOR IN UNEQUAL PROBABILITY SAMPLING

By

M. Samiuddin [(1)], Muhammad Hanif [(2)], A.K.A. Kattan [(3)] and Israr Khan[(4)]

SUMMARY

Godambe and Joshi (1965) derived the lower bound to expected variance of an estimator. This lower bound is rarely, if ever, achieved. This is because the class of estimators considered is very wide. In this paper we propose to limit the class of estimators which are both design- and model- biased. In section 2 of this paper an exact lower bound has been derived, whereas in section 3, an approximation to the lower bound has been derived using the Brewer's (1979) concept of asymptotic unbiasedness. Finally, empirical studies are presented.

Key words: Godambe- Joshi lower bound; design- and model-biased estimator; asymptotic unbiasedness.

1

[1] King Abdul Aziz University, Jeddah , [2] King Faisal University, Dammam, [3] Ummul-Qurah University, Makkah and [4] King Fahd University of Petroleum and Minerals, Dhahran, Saudi Arabia

1 INTRODUCTION

Consider a population of N units labelled $I = 1, 2, 3, \ldots, N$. Let Y_I be the characteristic of interest of the I-th unit in the population. Let Z_I be some characteristic for unit I which is known for all I. A random sample s of size n is drawn with probability $P(s)$ so that $\sum P(s) = 1$. The characteristic y_i $(i = 1, 2, 3, \ldots, n)$ is then known for all the sample units which leaves $N - n$ unsampled units with their Y_I's still unknown. It is required to estimate $Y = \sum_{I=1}^{N} Y_I$, using $(y_1, y_2, y_3, \ldots, y_n)$ and $Z_I(I = 1, 2, 3, \ldots, N)$. The model relating Y_I to Z_I is given by

$$Y_I = \beta Z_I + \epsilon_I, \quad (I = 1, 2, 3, \ldots, N) \tag{1.1}$$

where $E(\epsilon_I) = 0$ $\quad$ $\text{Var}(\epsilon_I) = \sigma_I^2$ and $\text{Cov}(\epsilon_I \epsilon_J) = 0$ for $J \neq I$.

Consider a general linear estimator of Y of the type

$$y'_s = \sum_{I \in s} C_{Is} Y_I. \tag{1.2}$$

Now y'_s is called design-unbiased if

$$E_D(y'_s) = \sum_s y'_s P(s) = Y$$

and is called model-unbiased if

$$E_M(y'_s) = Y. \tag{1.3}$$

If y'_s is design-unbiased then

$$\sum_{s \ni I} C_{Is} P(s) = 1, \qquad \text{for all } I \tag{1.4}$$

and if y'_s is model-unbiased then

$$\sum_{I \in s} C_{Is} Z_I = \sum_{I=1}^{N} Z_I = Z, \qquad \text{for all } s. \tag{1.5}$$

Godambe and Joshi (1965) derived a lower bound to the expected variance or *anticipated variance* (Isaki and Fuller 1982) of estimators of the type (1.2). This is given by

$$E_D E_M (y'_s - Y)^2 \geq \sum_{I=1}^{N} \sigma_I^2 \left(\frac{1}{\pi_I} - 1 \right) = V_1 \quad \text{(say)}, \tag{1.6}$$

where $\pi_I = \sum_{s \ni I} P(s)$.

In fact, this lower bound is rarely, if ever, achieved. This is because the class of estimators considered is very wide. A natural course of action is to limit the class of estimators and then look for the optimum estimator and the lower bound within this class. In this paper we propose to limit the class of estimators to those estimators of the form in (1.2) which are design- and model-unbiased, i.e., those which satisfy (1.4) and (1.5) simultaneously.

Many of the commonly used estimators satisfy both conditions. Some estimators are asymptotically design-unbiased; that is, these satisfy (1.4) approximately for large n [see Brewer (1979), Cassel, Sarndal and Wretman (1976)]. We shall see that subject to (1.4) and (1.5), we can, given $P(s)$, determine the best estimator y'_s, and also the lower bound to $E_M E_D (y'_s - Y)^2$ which is achievable.

2 A DESIGN- AND MODEL-UNBIASED ESTIMATOR

We assume y'_s to be both design- and model-unbiased. Its *anticipated variance* is

$$E_M E_D (y'_s - Y)^2 = E_M E_D \left[\sum_{I \in s} C_{Is} Y_I - \sum_{I=1}^{N} Y_I \right]^2 .$$

This reduces after some simplifications to

$$E_M E_D\,[y_s' - Y] = \sum_{I=1}^{N} \sigma_I^2 \sum_{s \ni I} C_{Is}^2 P(s) - \sum_{I=1}^{N} \sigma_I^2. \tag{2.1}$$

Now the design-expectation over the set of samples $s \ni I$ is

$$E_D(C_{Is}|s \ni I) = \sum_{s \ni I} C_{Is} \frac{P(s)}{\pi_I} = \frac{1}{\pi_I},$$

and the corresponding design-variance is

$$\mathrm{Var}_D(C_{Is}|s \ni I) = \sum_{s \ni I} C_{Is}^2 \frac{P(s)}{\pi_I} - \left(\frac{1}{\pi_I}\right)^2 \geq 0.$$

This leads to $\sum_{s \ni I} C_{Is}^2 P(s) \geq \frac{1}{\pi_I}$ and consequently

$$E_M E_D(y_s' - Y)^2 \geq \sum_{I=1}^{N} \sigma_I^2 \left(\frac{1}{\pi_I} - 1\right). \tag{2.2}$$

Equality in (2.2) holds iff $\mathrm{Var}_D(C_{Is}|s \ni I) = 0$, or $C_{Is} = K_I$ for all I, where K_I is a constant over all $s \ni I$. Now since $\sum_{I \in s} C_{Is} Z_I = Z$, we get $K_I (nZ_I/Z)^{-1}$. Also since $\sum_{s \ni I} C_{Is} P(s) = 1$, we get $\pi_I = nZ_I/Z$. Thus the Godambe–Joshi lower bound is achieved only if $\pi_I = nZ_I/Z$. Since $\sum_{I=1}^{N} \sigma_I^2 \left(\frac{1}{\pi_I} - 1\right)$ is minimized when $\pi_I = n\sigma_I / \sum_{I=1}^{N} \sigma_I$ (or $\pi_I \propto \sigma_I$) and also $\pi_I \propto Z_I$ an optimum strategy (choice of π_I as also C_{Is}) which achieves the bound given by Godambe–Joshi (1965) is available only for the case where $\sigma_I \propto Z_I$.

It is possible to choose C_{Is} in such a way that (1.4) and (1.5) are satisfied and $E_D E_M (y_s' - Y)^2$ is minimized. This leads to

$$C_{Is} = \frac{\mu_I}{\sigma_I^2} + \frac{\lambda_s}{P(s)} \frac{Z_I}{\sigma_I^2}, \tag{2.3}$$

where μ_I's and λ_s's are Lagrangian multipliers. Multiplying (2.3) by Z_I and summing over $I \in s$ gives

$$\sum_{I \in s} C_{Is} Z_I = \sum_{I \in s} \frac{\mu_I Z_I}{\sigma_I^2} + \frac{\lambda_s}{P(s)} \sum_{I \in s} \frac{Z_I}{\sigma_I^2} = Z.$$

This leads to

$$\frac{\lambda_s}{P(s)} = \frac{Z - \sum\limits_{I \in s} \mu_I Z_I / \sigma_I^2}{\sum\limits_{I \in s} Z_I^2 / \sigma_I^2}. \tag{2.4}$$

Using (2.4) in (2.3) gives

$$\begin{aligned} C_{Is} &= \frac{\mu_I}{\sigma_I^2} + \frac{Z - \sum\limits_{J \in s} \mu_J Z_J / \sigma_J^2}{\sum\limits_{J \in s} Z_J^2 / \sigma_J^2} \frac{Z_I}{\sigma_I^2} \\ &= C_I + \frac{Z - \sum\limits_{J \in s} C_J Z_J}{\sum\limits_{J \in s} Z_J^2 / \sigma_J^2} \frac{Z_I}{\sigma_I^2}, \quad \text{where } C_J = \frac{\mu_J}{\sigma_J^2}. \end{aligned} \tag{2.5}$$

C_I's can be solved iteratively using (1.4) which takes the new form

$$C_I = \frac{1}{\pi_I} - \frac{Z_I}{\sigma_I^2} \sum_{s \ni I} \left[\frac{Z - \sum\limits_{J \in s} C_J Z_J}{\sum\limits_{J \in s} Z_J^2 / \sigma_J^2} \frac{P(s)}{\pi_I} \right]. \tag{2.6}$$

For large sample sizes C_I is necessarily close to π_I^{-1}, since then $\sum\limits_{I \in s} C_I Z_I$ is the Horvitz–Thompson estimator for Z and approaches it asymptotically. Further, for numerous small sample cases that we have solved numerically C_I is indeed very close to π_I^{-1}. At any rate this provides a good first approximation to C_I. If we write $C_I^{(r)}$ as the r–th approximation to C_I (when $C_I^{(1)} = \pi_I^{-1}$), we get

$$C_I^{(r+1)} = \frac{1}{\pi_I} - \frac{Z_I}{\sigma_I^2} \sum_{s \ni I} \left[\frac{Z - \sum\limits_{I \ni s} C_J^{(r)} Z_J}{\sum\limits_{I \in s} Z_J^2 / \sigma_J^2} \frac{P(s)}{\pi_I} \right]. \tag{2.7}$$

Usually it took 2 to 3 iterations to get to the solutions in the examples that we worked out.

To simplify the expression for $E_M E_D (y_s' - Y)^2$ in (2.1) we proceed as follows. Multiplying (2.5) by $C_{Is} P(s)$ and summing over $s \ni I$ we get

$$\sum_{s \in I} C_{Is}^2 P(s) = C_I \sum_{s \ni I} C_{Is} P(s) + \sum_{s \ni I} \left[\frac{Z - \sum\limits_{J \in s} C_J Z_J}{\sum\limits_{J \in s} Z_j^2 / \sigma_J^2} C_{Is} P(s) \right] \frac{Z_I}{\sigma_I^2}$$

$$= C_I + \frac{Z_I}{\sigma_I^2} \sum_{s \ni I} \left[\frac{Z - \sum_{J \in s} C_J Z_J}{\sum_{J \in s} Z_j^2 / \sigma_J^2} C_{Is} P(s) \right]$$

Multiplying the above by σ_I^2 and summing over all I

$$\begin{aligned} \sum_{I=1}^{N} \left[\sigma_I^2 \sum_{s \ni I} C_{Is}^2 P(s) \right] &= \sum_{I=1}^{N} C_I \sigma_I^2 + \sum_{I=1}^{N} \left[Z_I \sum_{s \ni I} \frac{Z - \sum_{J \in s} C_J Z_J}{\sum_{J \in s} Z_J^2 / \sigma_J^2} C_{Is} P(s) \right] \\ &= \sum_{I=1}^{N} C_I \sigma_I^2 + Z \sum_{s} \left[\frac{Z - \sum_{J \in s} C_J Z_J}{\sum_{J \in s} Z_J^2 / \sigma_J^2} P(s) \right] \end{aligned} \quad (2.8)$$

Now $y_s' = \sum_{I \in s} C_{Is} Y_I$ and $E_D(y_s') = \sum_{I=1}^{N} Y_I$ for all Y_I's. Putting the value of C_{Is} we have

$$\begin{aligned} y_s' &= \sum_{I \in s} C_I Y_I + \frac{\sum_{I \in s} Z_I Y_I / \sigma_I^2}{\sum_{I \in s} Z_I^2 / \sigma_I^2} \left(Z - \sum_{I \in s} C_I Z_I \right) \\ E_D(y_s') &= \sum_{I=1}^{N} C_I Y_I \pi_I + \sum_{s} \left[\frac{Z - \sum_{I \in s} C_I Z_I}{\sum_{I \in s} Z_I^2 / \sigma_I^2} \sum_{I \in s} \frac{Z_I Y_I}{\sigma_I^2} P(s) \right] = \sum_{I=1}^{N} Y_I. \end{aligned}$$

Putting $Y_I = \sigma_I^2 / Z_I$ in the above expression

$$\begin{aligned} \sum_{I=1}^{N} C_I \pi_I \frac{\sigma_I^2}{Z_I} + \sum_{s} \left[\frac{Z - \sum_{I \in s} C_I Z_i}{\sum_{I \in s} Z_I^2 / \sigma_I^2} n P(s) \right] &= \sum_{I=1}^{N} \frac{\sigma_I^2}{Z_I} \\ \Rightarrow \quad \sum_{s} \left[\frac{Z - \sum_{I \in s} C_I Z_I}{\sum_{I \in s} Z_I^2 / \sigma_I^2} P(s) \right] &= \frac{1}{n} \left[\sum_{I=1}^{N} \frac{\sigma_I^2}{Z_I} - \sum_{I=1}^{N} C_I \pi_I \frac{\sigma_I^2}{Z_I} \right] \end{aligned} \quad (2.9)$$

Using (2.9) in (2.8) we get

$$\sum_{I=1}^{N} \sigma_I^2 \sum_{s \ni I} C_{Is}^2 P(s) = \sum_{I=1}^{N} C_I \sigma_I^2 + \frac{Z}{n} \left[\sum_{I=1}^{N} \frac{\sigma_I^2}{Z_I} - \sum_{I=1}^{N} C_I \pi_I \frac{\sigma_I^2}{Z_I} \right] \quad (2.10)$$

So (2.1) simplifies to

$$E_M E_D (y_s' - Y)^2 = \sum_{I=1}^{N} C_I \sigma_I^2 + \frac{Z}{n} \left[\sum_{I=1}^{N} \frac{\sigma_I^2}{Z_I} - \sum_{I=1}^{N} C_I \pi_I \frac{\sigma_I^2}{Z_I} \right] - \sum_{I=1}^{N} \sigma_I^2 = V_2 \quad \text{(say)} \quad (2.11)$$

provided C_I, and y'_s are chosen in an optimum way. Consequently, (2.11) provides the lower bound (achievable) for a given $P(s)$ to $E_M E_D(y'_s - Y)^2$. Expression (2.11) provides an exact result so that V_2 provides the lower bound for a given $P(s)$.

It would be useful to work out a case where the solution for C_I is explicit. In the Lahiri–Midzuno scheme (see Lahiri - 1951 and Midzuno - 1952), we set $P(s) \alpha \sum_{I \in s} Z_I^2/\sigma_I^2$ We then have $C_I = \dfrac{N-1}{n-1} - \dfrac{Z_I}{\sigma_I^2} A$, where A is any constant. It does not really matter what value of A is chosen since y'_s is always independent of A. Setting $A = 0$ we can obtain $C_I = (N-1)/(n-1)$ and π_I works out to $\pi_I = \dfrac{n-1}{N-1} + \dfrac{N-n}{N-1} \dfrac{Z_I^2}{\sigma_I^2} \left(\sum_{I=1}^{N} Z_I^2/\sigma_I^2 \right)^{-1}$ Now $E_M E_D(y'_s - Y)^2$ works out to be $\dfrac{N-1}{n-1} \left[\sum_{I=1}^{N} \sigma_I^2 - \dfrac{Z^2}{n} \left(\sum_{I=1}^{N} Z_I^2/\sigma_I^2 \right)^{-1} \right]$ which is the exact lower bound to $E_M E_D(y'_s - Y)^2$ for a given $P(s)$ and is achievable.

3 APPROXIMATION TO THE LOWER BOUND

Equation (2.11) gives lower bound to $E_M E_D(y'_s - Y)^2$ for estimators which are simultaneously design- and model-unbiased for a given $P(s)$. This however requires the solution for C_I's. Except in special cases this is not available in explicit form and has to be solved iteratively. We now indicate an approximation to the lower bound at (2.11). Put

$$W_I = Z_I/\sigma_I^2, \quad W_s = \sum_{J \in s} W_J Z_J, \quad A_s = \left(Z - \sum_{J \in s} C_J Z_j \right) W_S^{-1}$$

and $B_s = W_s^{-1} \sum_{J \in s} W_J E_D(A_s | s \ni J)$. From (2.6) C_I is given by

$$C_I = \frac{1}{\pi_I} - W_I E_D(A_s | s \ni I), \tag{3.1}$$

where

$$E_D(A_s | s \ni I) = \sum_{s \ni I} A_s \frac{P(s)}{\pi_I}$$

and

$$E_M E_D(y'_s - Y)^2 = \sum_{I=1}^{N} \sigma_I^2 \left(\frac{1}{\pi_I} - 1\right) + \sum_{I=1}^{N} W_I \sum_{s \ni I} [A_s - E_D(A_s|s \ni I]^2 P(s) \qquad (3.2)$$

Writing $A_s - E_D(A_s|s \ni I) = A_s - B_s + B_s - E_D(A_s|s \ni I)$ the second term on R.H.S. of (3.2) becomes

$$\sum_{I=1}^{N} W_I \sum_{s \in I} \left[(A_s - B_s)^2 + \{B_s - E_D(A_s|s \ni I)\}^2 + 2(A_s - B_s)(B_s - E_D(A_s|s \ni I\right] P(s).$$

The product term vanishes by writing $\sum_{I=1}^{N} \sum_{s \ni I}$ in the form of $\sum_s$ because

$$\begin{aligned}
&\sum_{I=1}^{N} W_I \sum_{s \in I} [A_s - B_s][B_s - E_D(A_s|s \ni I)] P(s) \\
&= \sum_s (A_s - B_s) \sum_{I \in s} \left[B_s \sum_{I \in s} W_I - \sum_{I \in s} W_I E_D(A_s|s \ni I)\right] P(s) \\
&= 0,
\end{aligned}$$

since $B_s \sum_{I \in s} W_I = \sum_{I \in s} W_I E_D(A_s|s \ni I)$ by definition.

So

$$E_M E_D(y'_s - Y)^2 = \sum_{I=1}^{N} \sigma_I^2 \left(\frac{1}{\pi_I} - 1\right) + \sum_{I=1}^{N} W_I \sum_{s \ni I} \left[(A_s - B_s)^2 + \{B_s - E_D(A_s|s \ni I)\}^2\right] P(s) \qquad (3.3)$$

Multiplying both sides of (3.1) by Z_I, summing over $I \in s$ and rearranging term we get

$$\sum_{I \in s} C_I Z_I = \sum_{I \in s} \frac{Z_I}{\pi_I} - \sum \frac{Z_I^2}{\sigma_I^2} E_D(A_s|s \ni I)$$

and

$$Z - \sum_{I \in s} C_I Z_I = \left(Z - \sum_{I \in s} \frac{Z_I}{\pi_I}\right) + \sum_{I \in s} W_I E_D(A_s|s \ni I)$$

so that

$$\left(Z - \sum_{I \in s} C_I Z_I\right) W_s^{-1} = \left(Z - \sum_{I \in s} \frac{Z_I}{\pi_I}\right) W_s^{-1} + B_s$$

This finally leads to

$$A_s - B_s = \left(Z - \sum_{I\in s}\frac{Z_I}{\pi_I}\right) W_s^{-1}$$

From (3.3), we get

$$E_M E_D(y_s' - Y)^2 \geq \sum_{I=1}^{N} \sigma_I^2\left(\frac{1}{\pi_I} - 1\right) + \sum_s \frac{\left(Z - \sum_{I\in s}\frac{Z_I}{\pi_I}\right)^2}{\sum_{I\in s} Z_I^2/\sigma_I^2} P(s) \tag{3.4}$$

Equality holds iff $E_D(A_s|s \ni I) = B_s$. using Brewer's (1979) type approximation

$$\sum_s \frac{\left(Z - \sum_{I\in s}\frac{Z_I}{\pi_I}\right)^2}{\sum_{I\in s} Z_I^2/\sigma_I^2} P(s) \simeq \frac{\sum_s \left(Z - \sum_{I\in s}\frac{Z_I}{\pi_I}\right)^2 P(s)}{\sum_s\sum_{I\in s}\frac{Z_I^2}{\sigma_I^2}P(s)} = \frac{\text{Var}(z_{HT}')}{\sum_{I=1}^{N}\frac{Z_I^2}{\sigma_I^2}\pi_I}$$

Consequently we may expect for n, $N \to \infty$

$$E_M E_D(y_s' - Y)^2 \geq \sum_{I=1}^{N} \sigma_I^2\left(\frac{1}{\pi_I} - 1\right) + \frac{\text{Var}(z_{HT}')}{\sum_{I=1}^{N}\frac{Z_I^2}{\sigma_I^2}\pi_I} = V_3 \quad \text{(say)}. \tag{3.5}$$

It is interesting to note that one part of the bound is minimized if $\pi_I \propto \sigma_I$ and the other part is minimized if $\pi_I \propto Z_I$ (when this part becomes zero). Optimum choice of π_I would thus seem to lie between the two extreme cases. The numerator of the second part is the variance of the Horvitz and Thompson estimator which for large n will be small. This suggests that the optimum choice of the π_I would be near to the one suggested by Brewer (1979), namely $\pi_I \propto \sigma_I$. This is borne out by the numerical study in the next section.

4 NUMERICAL COMPARISON

The calculation of V_1, V_2 and V_3, defined earlier, depends on the values of Z_I, σ_I^2 and C_I, π_J and π_{IJ} [which in turn depends on $P(s)$]. We considered the populations given in Baylass and Rao (1970) and others from different textbooks. This was done so that Z_I values correspond to quite a number of real populations. We have set $\sigma_I^2 = Z_I^{2\gamma}$ and 4 different values of γ are selected. These are $\gamma = 0.50, 0.625, 0.750$ and 0.875. We have dropped $\gamma = 1$ because in this case the optimum strategy is completely known. We have then selected 4 sets of values of $\pi_I \propto Z_I$ and the values selected for δ where $\delta = 0.50, 0.625, 0.750, 0.875$. So, for each of these populations we had 16 different cases corresponding to pairs of values of (γ, δ). We have chosen the Samiuddin–Asad (1981) selection procedure because it gives π_I and π_{IJ} in explicit form quite easily for all n. The result presented here corresponds to a populations of 10 and 20 units with Z_I values given by $Z_I = 60, 52, 58, 56, 62, 51, 72, 48, 71, 58$. [Cochran (1977)] and $Z_I = 18, 9, 14, 12, 24, 25, 23, 24, 17, 14, 18, 40, 12, 30, 27, 26, 21, 9, 19, 12$. (Horvitz and Thompson, 1952) and $n = 2$ and 4 (in Tables 1a, 1b and 2a, 2b) respectively. The results for other populations parallel those provided here, and lead to very similar conclusions. Very interesting conclusions emerge from these calculations. However, we must bear in mind that we are using a sampling scheme which would be considered good.

The first important conclusion is that Godambe–Joshi lower bound provides a rather tight bound even for n as low as 2. One need only to compare V_1 (Godambe–Joshi bound) with V_2 (the exact achievable lower bound) for each pair of (γ, δ). The second important conclusion is that choice of π_I reflected by the choice of δ has practically no effect on $E_M E_d(y_s' - Y)^2$, given the selection procedure in Samiuddin and Asad (1981). To compare the values of V_1 and V_2 in the same row, notice that V_1 is always less than V_2 as it should be. A third point of some interest is that the approximate lower bound V_3 is always between V_1 and V_2. Note that the lowest value of V_2 does not occur when $\gamma = \delta$ in case $n = 2$ but the lowest value of V_2 does occur when $\gamma = \delta$ as Brewer (1979) suggests in case $n = 4$. In fact for $n = 2$, it matches the closing remarks in section 3.

Table 2(a): Values of V_1, V_2 and V_3 for different pairs of γ and δ for $n = 2$.

$\gamma\backslash\delta$		.500	.625	.750	.875
	V_1	3400	3409	3436	3481
.500	V_2	3487	3473	3478	3502
	V_3	3444	3434	3447	3484
	V_1	7183	7165	7163	7239
.625	V_2	7326	7270	7252	7276
	V_3	7279	7218	7227	7273
	V_1	15315	15198	15159	15198
.750	V_2	15519	15351	15260	15248
	V_3	15515	15313	15211	15220
	V_1	32944	32531	32285	32203
.875	V_2	33370	32776	32401	32259
	V_3	33187	32710	32397	32232

Table 2(b): Values of V_1, V_2 and V_3 for different pairs of γ and δ for $n = 4$.

$\gamma\backslash\delta$		.500	.625	.750	.875
	V_1	1504	1508	1521	1543
.500	V_2	1527	1520	1526	1546
	V_3	1516	1511	1523	1544
	V_1	3167	3158	3167	3195
.625	V_2	3192	3172	3179	3201
	V_3	3185	3186	3172	3196
	V_1	6734	6676	6656	6676
.750	V_2	6773	6705	6675	6685
	V_3	6772	6698	6667	6679
	V_1	14449	14242	14119	14079
.875	V_2	14549	14300	14147	14091
	V_3	14513	14287	14145	14085

Table 3: Values of V_1, V_2 and V_3 for $n = 2$ using Rao–Sampford selection procedure.

Pop	V	.500	.625	.750	.875
	V_1	2340.2	6478.2	17939.8	49698.4
1	V_1	2350.6	6494.3	17960.0	49712.0
	V_3	2345.5	6486.4	17950.4	49705.0
	V_1	3400.6	7165.3	15160.0	32204.0
2	V_1	3535.7	7320.4	15320.0	32278.0
	V_3	3462.7	7239.4	15230.0	32242.0
	V_1	2257.8	6197.7	17023.0	46784.5
2	V_1	2273.7	6222.2	17053.0	46805.0
	V_3	2265.9	6210.3	17038.4	46795.0
	V_1	592.4	814.05	1127.9	1572.9
2	V_1	631.7	842.90	1144.7	1578.7
	V_3	608.7	825.30	1134.4	1575.0

Table 4: Values of V_1, V_2 and V_3 for $n = 4$ using Rao–Sampford selection procedure.

Pop	V	.500	.625	.750	.875
	V_1	876.1	2422.9	6701.7	18539.0
1	V_1	878.0	2425.7	6705.3	18542.0
	V_3	877.1	2424.4	6703.6	18541.0
	V_1	1503.3	3158.0	6656.9	14079.0
2	V_1	1529.7	3188.8	6685.6	14094.0
	V_3	1516.9	3174.3	6672.2	14087.0
	V_1	844.4	2314.3	6344.5	17397.0
2	V_1	847.3	2318.7	6349.8	17401.0
	V_3	845.9	2316.6	6347.3	17399.0
	V_1	260.7	356.6	491.0	680.1
2	V_1	267.9	362.1	494.4	681.3
	V_3	264.3	359.4	492.7	680.7

Finally, the conclusions for all other cases remain almost exactly as in the case for Tables 1a, 1b and 2a, 2b.

We also performed some comparisons using Rao–Sampford selection procedure. This selection procedure is also supposed to be good in performance terms. The calculations here are not as comprehensive as in the case of Tables 1 and 2. The reason for doing this was basically to check if the conclusions drawn are dependent on the type of selection procedure chosen. We again observe that V_1, V_2 and V_3 are close together, V_1 is less than V_2. The results are presented in Tables 3 and 4 for $n = 2$ and $n = 4$, respectively. In all calculations, we have set $\sigma_I^2 = Z_I^{2\gamma}$ when $\gamma = 0.5, 0.625, 0.750, 0.875$ as before and $\pi_I \propto \sigma_I$. The result for 4 populations are provided in Tables 3 and 4 in which populations 1 and 2 are the same as in Tables 1(a), 1(b) and 2(a), 2(b). In actual calculation, first the values of C_i's are determined by the iteration process. These are then used to calculate V_2. The calculation of V_3 involves the calculation of Var(z'_{IIT}). The values of other two populations are

Pop. 3: Z_I = 59, 47, 52, 60, 67, 48, 44, 58, 76 and 58 (Cochran, 1977)

Pop. 4: Z_I = 6, 5, 4, 5, 4, 2, 4, 2, 5, 1, 3, 4, 3, 1, 1, 3, 4, 8, 2, and 4 (Sukhatme and Sukhatme, 1970).

ACKNOWLEDGEMENT

We are grateful to Prof. K.R.W. Brewer of the Australian National University, Canberra, Australia for his useful comments which helped to improve the text of this paper.

REFERENCES

1. Bayless, D.L. and Rao, J.J.K. (1970). An empirical study of the stabilities of estimators and variance estimators in unequal probability sampling ($n = 3, 4$). *J. Amer. Statist. Assoc.* 65, 1645–1667.

2. Brewer, K.R.W. (1979). A class of robust sampling designs for large scale surveys. *J. Amer. Statist. Assoc.* 74(4), 911–915.

3. Cassel, Class–M, Sarndal, C–E. and Wretman, Jan–H (1976). Some results on generalized estimation and generalized regression estimation for finite populations. *Biometrika*, 63, 615–620.

4. Cochran, W.G. (1977). *Sampling Techniques*, John Wiley, U.S.A.

5. Godambe, V.P. and Joshi, V.M. (1965). Admissibility and Bayes estimation in sampling finite populations, I, II and III. *Ann. Math. Statist.* 36, 1707–1742.

6. Horvitz, D.G. and Thompson, D.J (1952). A generalization of sampling without replacement from a finite universe. *J. Amer. Statist. Assoc.* 47, 663–685.

7. Isaki, C.T. and Fuller, W.A. (1982). Survey design under the regression superpopulation model. *J. Amer. Statist. Assoc.* 77, 209–218.

8. Lahiri, D.B. (1951). A method of sample selection providing unbiased ratio estimates, *Bull. Inter. Statist. Inst XXXIII Book 2*, 133–140.

9. Midzuno H. (1952). On the sampling system with probability proportional to sum of sizes. *Ann. Inst. Statist. Math.* 3, 99–107.

10. Rao, J.N.K. (1965). On two simple schemes of unequal probability sampling without replacement. *J. Indian Statist. Assoc.* 3, 99–107.

11. Samiuddin, M. and Asad, H. (1981). A simple procedure of unequal probability sampling. *Biometrika*, 68(3), 728–731.

12. Sampford, M.R. (1967). On sampling without replacement with unequal probabilities of selection. *Biometrika*, 54, 499–513.

13. Sukhatme, P.V. and Sukhatme, B.V. (1970). *Sampling theory of surveys with applications.* 2nd Edition, Asia Publishing House, India.

Applied Statistical Science, II
ISBN 1-56072-469-2

ASYMPTOTIC SHRINKAGE ESTIMATION: THE REGRESSION CASE

S. E. Ahmed

Department of Mathematics and Statistics
University of Regina
Regina, Saskatchewan Canada S4S 0A2
Internet: ahmed@albiruni.math.uregina.ca

January 13, 1997

Abstract

In this paper, we first outline some recent developments in the area of pretest and shrinkage estimation. We discuss various large sample estimation techniques in a regression model when the error terms are not necessarily normally distributed. The commonly asked question whether to combine sample data and *nonsample information (NSI)* or *uncertain prior information (UPI)* will be systematically addressed. We propose estimators on the basis of preliminary tests of significance and James-Stein rule. The properties of these estimators are studied in the problem of estimating regression coefficients in the multiple regression model when it is apriori suspected that the coefficients may be restricted to a subspace.

1. INTRODUCTION

Statistical models are estimated in an effort to have insights about unknown quantities (parameters). In many situations, however, the statisticians provide the estimation of the parameters by using information based on the sample data and other information as well. The other information may be regarded as *nonsample information(NSI)* or *uncertain prior information (UPI)* about the parameter of interest. It is advantageous to utilize the NSI in the estimation procedure, specially when the information based on the sample may be rather limited.

The use of reliable information usually gives fruitful results. But in some experimental cases, it is not certain whether or not this information hold. Consider the data arising from tumor measurements in mice, for example, at various times following injection of carcinogens. Such data should be thought as coming from an *in vivo* experiment. Biologists are interested in estimating growth rate parameter λ when it suspected *apriori* that $\lambda = \lambda_o$. Such a λ_o can be obtained directly from *in virto* experiments in which cell behavior may or may not be different because of different environmental conditions. Thus, biologists may have reason to suspect that λ_o is the true value of the growth parameter for the *in vivo* experiment, but may not sure. In fact, in the applied sciences, an experiment is often performed with some knowledge of the likely outcome of the experiment. The use of standard statistical rules, which usually assume complete ignorance about the outcome, may not be useful.

Generally speaking, consequences of incorporating nonsample information depend on the *quality or reliability* of information introduced in the estimation process. This uncertain prior information, in form of the null hypothesis, can be used in two different ways in the estimation procedure. At first place, it is natural to perform a preliminary test on the validity of the *UPI* in the form of parametric restrictions, and then choose between the restricted and unrestricted estimation procedure depending upon the outcome of the preliminary test. This idea was initially conceived by T. A. Bancroft in 1944. However, this may be partly motivated by the remarks made by Berkson (1942).

In the later case, the James-Stein estimation procedure is adopted and we intend to discuss this case first. For the past three decades, the researchers have paid considerable attention to the James-Stein type estimation, in small sample as well as in large sample set ups. Readers may find a sufficient amount of theoretical developments in the current literature.

1.1. Statement of the Problem

To this end, for expository purposes, let us formulate the basic problem of estimating mean vector parameter of a multivariate normal distribution. To do so, let $\mathbf{x} = (x_1, x_2, \cdots, x_k)$ be a random vector having k-variate normal distribution with mean parameter vector $\boldsymbol{\theta} = (\theta_1, \cdots, \theta_k)'$ and covariance matrix $\mathbf{I}_k$ and the parameter of interest is $\boldsymbol{\theta}$ which need to be estimated statistically.

1.2. Shrinkage Estimation

Stein (1956) has demonstrated that the *maximum likelihood (ml)* estimator of $\boldsymbol{\theta}$ is inadmissible for $k > 2$ when a quadratic loss function is considered. It is important to note that the inadmissibility does not hold for $k = 1$ and $k = 2$.

1.2.1. Shrinking $\hat{\boldsymbol{\theta}}^U$ towards the null vector

James and Stein (1961) presented an explicit form of an estimator which dominates the usual ml estimator for $k \geq 3$. Let $\hat{\boldsymbol{\theta}}^U = (\hat{\theta}_1^U, \cdots, \hat{\theta}_k^U)'$ be the usual ml estimator of $\boldsymbol{\theta}$ then a simple form of the James-Stein type estimator of $\boldsymbol{\theta}$ is then given by

$$\hat{\boldsymbol{\theta}}_1^S = \left\{1 - \frac{c}{(\hat{\boldsymbol{\theta}}^U)'(\hat{\boldsymbol{\theta}}^U)}\right\}\hat{\boldsymbol{\theta}}^U, \tag{1.1}$$

where c is a constant, generally known as the the shrinkage constant, such that $0 \leq c \leq 2(k-2)$. This estimator has smaller risk than ml estimator for all valúes of $\boldsymbol{\theta}$ such that $\boldsymbol{\theta}'\boldsymbol{\theta} < \infty$ if $k \geq 3$. When the mean vector $\boldsymbol{\theta}$ is a null vector, the optimal estimator (1.2) offers the minimum risk which increases towards the risk of the ml estimator as $\boldsymbol{\theta}'\boldsymbol{\theta} \to \infty$. We notice that $\hat{\boldsymbol{\theta}}_1^S$ shrinks $\hat{\boldsymbol{\theta}}^U$ towards the origin and such an estimator is generally called a *shrinkage estimator (SE)*. However, there is no reason why $\hat{\boldsymbol{\theta}}^U$ must shrink towards the origin. In general we may shrink $\hat{\boldsymbol{\theta}}^U$ towards any arbitrary fixed vector $\boldsymbol{\vartheta}$.

1.2.2. Shrinking $\hat{\boldsymbol{\theta}}^U$ towards a hypothesis vector

If one wish to shrink $\hat{\boldsymbol{\theta}}^U$ towards a hypothesized vector $\boldsymbol{\vartheta}$ then in this case the optimum James-Stein estimator is

$$\hat{\boldsymbol{\theta}}_2^S = \left\{1 - \frac{k-2}{(\hat{\boldsymbol{\theta}}^U - \boldsymbol{\vartheta})'(\hat{\boldsymbol{\theta}}^U - \boldsymbol{\vartheta})}\right\}(\hat{\boldsymbol{\theta}}^U - \boldsymbol{\vartheta}) + \boldsymbol{\vartheta}, \quad k \geq 3, \tag{1.2}$$

where $\boldsymbol{\vartheta}$ is an arbitrary fixed vector. In fact the *ml* estimator can be shrunk towards the general restricted *ml* estimator.

1.2.3. Shrinking $\hat{\boldsymbol{\theta}}^U$ towards a restricted estimator vector

The *ml* estimator may also be shrinked towards an estimator instead of a known quantity. As an example, Rao (1948) considered a 28×4 data matrix showing the weights of the bark deposits of 28 trees in the four directions, north, south, east and west. If we treat these trees to be a random sample from the population of all such trees and denote the weights of the bark deposits by x_{north}, x_{south}, x_{east} and x_{west}, then our sample will consist of 28 observations on the vector $\mathbf{x} = (x_{north}, x_{south}, x_{east}, x_{west})'$. Assume that $\mathbf{x}$ follows a normal distribution with mean vector $\boldsymbol{\theta} = (\theta_{north}, \theta_{south}, \theta_{east}, \theta_{west})'$ with a covariance matrix of order 4×4. Suppose that we are interested in the estimation of the mean vector of deposits. One possible hypothesis is that the mean bark deposit for the population is the same. A common use that is made of this test is the so-called *test of symmetry*. In other words, $H_o : \theta_{north} = \theta_{south} = \theta_{east} = \theta_{west} = \theta_o$ (*unknown*). In general, we may have $\theta_1 = \theta_2 = \cdots = \theta_k = \theta_o$ (*unknown*). In this case, we first define a *restricted estimator (RE)* denoted by $\hat{\boldsymbol{\theta}}^R$ as $\hat{\boldsymbol{\theta}}^R = (\hat{\theta}^R, \cdots, \hat{\theta}^R)' = \hat{\theta}^R \mathbf{1}_k$, $\mathbf{1}_k' = (1, \cdots, 1)$, where $\hat{\theta}^R = \frac{1}{k}\sum_{i=1}^k \bar{x}_i$. It is well documented in literature that restricted estimators are better than unrestricted ones when the scale difference between guessed value of parameter and actual value of the parameter is small. As this scale difference grows, $\hat{\boldsymbol{\theta}}^R$ is no longer a desirable estimator, since it may be considerably biased, inefficient and inconsistent, while the performance of the $\hat{\boldsymbol{\theta}}^U$ remains unaffected.

We may shrink $\hat{\boldsymbol{\theta}}^U$, the usual *ml* estimator of $\boldsymbol{\theta}$ towrds $\hat{\boldsymbol{\theta}}^R$. In this case shrinkage or James-Stein estimator is

$$\hat{\boldsymbol{\theta}}_3^S = \hat{\boldsymbol{\theta}}^R + \left\{1 - \frac{c}{(\hat{\boldsymbol{\theta}}^U - \hat{\boldsymbol{\theta}}^R)'(\hat{\boldsymbol{\theta}}^U - \hat{\boldsymbol{\theta}}^R)}\right\}(\hat{\boldsymbol{\theta}}^U - \hat{\boldsymbol{\theta}}^R). \tag{1.3}$$

All the James-Stein estimators considered above are minimax and dominate the *ml* estimators in the entire parameter space regardless of the correctness of the *UPI*. It is important to remark here that the shrinkage estimators do not dominate restricted estimators.

We observe that James-Stein type estimator is nonlinear, biased and has rather nonintuitative nature. We also wish to remark here that, *ml* estimator is inadmissible with respect to James-Stein type estimator while estimating the several parameters simultaneously. On the other hand, the components of the *ml* estimator are separately admissible to estimate the corresponding one dimensional parameter. So the James-Stein type estimator dominates the *ml* estimator with respect to a loss function which sums the squared loss of estimation from each parameter. This fact is known as Stein effect or Stein paradox. Also, if we seek an estimator of $\boldsymbol{\theta}$ and adopt the weighted loss function, then James-Stein type domination over *ml* estimator will require additional design related conditions. In such a case, to induce the dominance of $\hat{\boldsymbol{\theta}}^S$ over $\hat{\boldsymbol{\theta}}^U$, one may need to introduce the modified James-Stein estimator and we refer to Berger et al. (1977) and Sen and Saleh (1985).

Further, there is an unpleasant feature of $\hat{\boldsymbol{\theta}}^S$ that the $\hat{\boldsymbol{\theta}}^U$ may shrunk beyond the null hypothesis or restricted estimator vector yielding the sign change of the *ml* estimator $\hat{\boldsymbol{\theta}}^U$. One may agree on adjusting the magnitudes of the $\hat{\boldsymbol{\theta}}^U$, but change of sign is somewhat an unpleasant property and it would make a practitioner rather uncomfortable in such a case. It is important to note that this behavior of $\hat{\boldsymbol{\theta}}^S$ may not adversely affect the risk. Fortunately, this disquieting feature i.e., over shrinking problem of $\hat{\boldsymbol{\theta}}^S$ can be rectified by truncating $\hat{\boldsymbol{\theta}}^S$ with its positive part.

1.2.4. *Positive-part Shrinkage Estimator*

The James-Stein estimators considered above have smaller risk than the *ml* estimators over the entire parameter space. However, they are inadmissible. That is, there are other rules which dominate the James-Stein estimators. The shrinkage estimator in (1.2) may be improved by the positive-part shrinkage estimators, and given by,

$$\hat{\boldsymbol{\theta}}_1^{S+} = \left\{1 - \frac{k-2}{(\hat{\boldsymbol{\theta}}^U)'(\hat{\boldsymbol{\theta}}^U)}\right\}^+ \hat{\boldsymbol{\theta}}^U, \quad k \geq 2, \tag{1.4}$$

where $a^+ = max\{0, a\}$. These estimators prevent changing the sign of $\hat{\boldsymbol{\theta}}^U$. Ahmed (1992a) and Ahmed and Khan (1996) provided an explicit form for risk of the positive-part estimator in a large sample set up. Hwang and Cassela (1982, 1984) have used the positive-part estimator to construct the confidence set for $\boldsymbol{\theta}$ in an explicit form. Their work has been extended by Hwang and Chen (1986), Chen and Hwang (1988) and Hwang and Ullah (1994) among others. While the positive-part shrinkage estimator has lower risk than $\hat{\boldsymbol{\theta}}^S$ but is inadmissible itself. To this end, Shao and Strawderman (1994) were able to obtain an estimator which dominates the positive-part shrinkage estimator. However, their estimator is also inadmissible, and there is further scope of research in this direction.

The ad hoc nature of the James-Stein estimator may have limited the application of their ideas at a large scale. However, intuition about how the James-Stein estimator work can greatly be enhanced by considering the example provided by Efron and Morris (1977). Efron and Morris (1972) and Robbins (1983) noted that the estimators arises in a natural fashion as an empirical Bayes estimator. Thus, the James-Stein estimator has a statistical base and is not pulled out of the hat. Brandwein and Strawderman (1990) and Stigler (1990) also provided interpretation of the James-Stein estimator. For an account of the parametric theory of Stein-type estimators, we refer to Judge and Bock (1978), Berger (1980), Anderson (1984), Ahmed and Saleh (1989, 1990, 1993), Robert (1994), and Rukhin (1995) among others. Asymptotic theory of some of these estimators have been studied by Sen (1986), Gupta et al. (1989), Ali and Saleh (1991) Ahmed (1992a, 1996a, 1996b), Ahmed and Khan (1996), Ahmed and Gupta (1996). The nonparametric asymptotic estimation theory is considered by Ahmed and Saleh (1996) and Saleh and Sen (1978-1985). Further, covariance matrix of the Stein-type estimators depend on the unknown population parameters. However, some progress have been made in this area and readers are referred to Haff (1979), Dasgupta (1989) among others.

1.3. Pretest Estimation

Recall that generally shrinkage estimators are superior to ml estimators for $k \geq 3$ while ml estimator is admissible for $k = 1$ and $k = 2$. Thus, the use of the shrinkage estimation may be limited due to dimensional restriction. As such, in bivariate case we will be unable to use NSI in the estimation process. In this situation, we recommend to use estimators based on pretest rule.

The *preliminary test/pretest estimators (PTE)* are widely used by researchers, as is evident from the extensive bibliographies of Bancroft and Han (1977), Han *et al.*(1988). However, the estimators generated by the preliminary test methodology are superior to the estimators based on sample data only in a relatively narrow portion of the parameter space induced by the nonsample information. The performance of the PTE depends on the quality of the UPI. If the information is correct or nearly correct then PTE is better than the estimator based on sample information while opposite conclusion holds, otherwise.

1.3.1. Standard Pretest Estimator

The pretest estimator for the parameter vector $\boldsymbol{\theta}$ which emerges from the classical pretest rule (Bancroft, 1944) is $\hat{\boldsymbol{\theta}}^P$, where

$$\hat{\boldsymbol{\theta}}^P = \hat{\boldsymbol{\theta}}^R I(D_n \leq d_{n,\alpha}) + \hat{\boldsymbol{\theta}}^U I(D_n > d_{n,\alpha}), \tag{1.5}$$

where D_n be an appropriate test statistic for the null hypothesis $H_o : \theta_1 = \cdots = \theta_k = \theta_o$. Further, $I(A)$ is the indicator function of the set A and $d_{n,\alpha}$, for $\alpha \in (0,1)$, is the upper α-percentile of the distribution of D_n under H_o.

In a preliminary test estimation problem, one sets to test the given NSI before choosing the estimator, while James-Stein type estimation incorporates the NSI in the estimator to adjust for the possible weight of the NSI. The shrinkage procedure may also be viewed as a smooth version of the Bancroft (1944) method. For an account of the parametric theory of pretest estimators in the finite sample space, we refer to Ahmed and Saleh (1989, 1990, 1993), Judge and Bock (1978), among others. Asymptotic theory of these estimators have been studied by Sen (1986), Ahmed (1991a, 1991b, 1992a), Gupta et. al (1989), Kulperger and Ahmed (1992) and others. The nonparametric asymptotic estimation of the pretest estimators is considered by Ahmed and Saleh (1996), Koul and Saleh (1994) and Saleh and Sen (1978 - 1985) among others.

We notice that $\hat{\boldsymbol{\theta}}^P$ is a function of sample data, hypothesis and α, the size of the pretest. The later feature, i.e., the size of the preliminary test is an important characteristic of the preliminary test estimation which is mostly ignored. However, it is recommended in the literature to use a level of significance of at least 0.15 or more for such preliminary testing. Use of such a large significance level helps maximize the minimum efficiency of $\hat{\boldsymbol{\theta}}^P$. The use of $\hat{\boldsymbol{\theta}}^P$ is limited due to the large size of the preliminary test. A shrinkage technique is therefore introduced into the pretest estimation to overcome this difficulty. The proposed methodology remarkably improves upon the PTE with respect to the size of the preliminary test (Ahmed, 1992a).

1.3.2. *Shrinkage Pretest Estimator*

We first define a *shrinkage restricted estimator (SRE)* of $\boldsymbol{\theta}$ as

$$\hat{\boldsymbol{\theta}}^{SR} = \pi\hat{\boldsymbol{\theta}}^{U} + (1-\pi)\hat{\boldsymbol{\theta}}^{R},$$

where π is a coefficient reflecting degree of distrust in the prior information. The value of π may be completely determined by the experimenter, depending upon the degree of his disbelief in the NSI. A value near 1 causes $\hat{\boldsymbol{\theta}}^{SR}$ to be based essentially on the sample data alone. Then *shrinkage preliminary test estimator (SPTE)* of $\boldsymbol{\theta}$ denoted by $\hat{\boldsymbol{\theta}}^{SP}$ is defined by replacing $\hat{\boldsymbol{\theta}}^{R}$ by $\hat{\boldsymbol{\theta}}^{SR}$ in $\hat{\boldsymbol{\theta}}^{P}$ to obtain

$$\hat{\boldsymbol{\theta}}^{SP} = \hat{\boldsymbol{\theta}}^{U} - (1-\pi)(\hat{\boldsymbol{\theta}}^{U} - \hat{\boldsymbol{\theta}}^{SR})I(D_n \leq d_{n,\alpha}). \tag{1.6}$$

For $\pi = 0$, $\hat{\boldsymbol{\theta}}^{SP} = \hat{\boldsymbol{\theta}}^{P}$ and for $\pi = 1$, $\hat{\boldsymbol{\theta}}^{SP} = \hat{\boldsymbol{\theta}}^{U}$. The properties of this estimator in various situations have been studied by Ahmed(1992 -1995), Ahmed and Rohatgi (1996), Ahmed and Braun (1996) among others.

However, the usual pretest estimator given in relation (1.10) may be improved by replacing $\hat{\boldsymbol{\theta}}^{U}$ with $\hat{\boldsymbol{\theta}}^{S}$, and such an estimator is generally called improved pretest estimator. This estimator was first introduced by Sclove et al. (1972).

1.3.3. *Improved Pretest Estimation*

It can be noted that $\hat{\boldsymbol{\theta}}^{P}$ may also be uniformly improved by inserting shrinkage estimator instead of $\hat{\boldsymbol{\theta}}^{U}$ in $\hat{\boldsymbol{\theta}}^{P}$. We call it *improved preliminary test estimator (IPTE)* and is given by the following relation

$$\hat{\boldsymbol{\theta}}^{P+} = \hat{\boldsymbol{\theta}}^{R}I(D_n \leq d_{n,\alpha}) + \hat{\boldsymbol{\theta}}^{S}I(D_n > d_{n,\alpha}). \tag{1.7}$$

This estimator dominates the standard pretest estimator but we will now generally have the restriction that $k \geq 2$. However, it is important to note that improved pretest estimator does not improve upon $\hat{\boldsymbol{\theta}}^{U}$ uniformly and hence embraces the similar kind of criticism as being absorbed by $\hat{\boldsymbol{\theta}}^{P}$. Ahmed and Ullah (1996a, 1996b) and others examined the properties of this estimator. The asymptotic behavior of improved pretest estimator in nonparamertic situation is discussed by Ahmed and Saleh (1996).

The pretest estimation literature mostly deals with point estimation of the parameter(s). However, the associated problem of finding confidence intervals based on the pretest estimator has not been studied much. The conditional interval estimation for a normal mean following a rejection of a pretest has been studied by Meek and D'Agostino (1983) and Arabatzis et al. (1983). Ahmed and Kulperger (1990) provided asymptotic confidence interval from a preliminary test estimator based on two samples taken from two arbitrary populations. The conditional confidence intervals for the exponential scale and location parameter, respectively, have been studied by Choi and Han (1994, 1995).

Now, we turn our attention to the main objective of the paper which is to focus on the large sample properties (under quadratic loss) of the estimators based on preliminary test and James-Stein rule estimators and to provide a comprehensive comparative study of these estimators. Specifically, we will consider estimation of the regression coefficients in the multiple regression model when it is apriori suspected that the coefficient may be restricted to a subspace. It will be shown that preliminary test estimators improve significantly upon the usual estimators in a region of the parameter space. The shrinkage estimators dominate the usual estimators, while shrinkage estimators are dominated by its truncated part. Not surprisingly, none of shrinkage and preliminary test estimators dominate each other.

2. Regression Analysis: Nonnormal Disturbances

Consider the linear regression model

$$\mathbf{y}_n = \mathbf{X}_n\boldsymbol{\beta} + \boldsymbol{\epsilon}_n, \tag{2.1}$$

$\mathbf{y}_n$ be an $n \times 1$ random vector, $\mathbf{X}_n$ is a known $n \times k$ matrix, $(n > k)$ of regression constant and, as the sample size n become infinitely large, $\lim_{n\to\infty}(\mathbf{X}_n'\mathbf{X}_n) = \mathbf{Q}$ where $\mathbf{Q}$ is finite and nonsingular matrix. Further, $\boldsymbol{\beta}$ is a column vector of k unknown regression parameters, $\epsilon = (\epsilon_1, \cdots, \epsilon_k)'$ are independent and identically distributed random variables with a distribution function F, on real line $\Re = (-\infty, +\infty)$. Here we do not make any assumption about the functional form of the F. We assume that $E(\epsilon) = \mathbf{0}$ and $E(\epsilon\epsilon') = \sigma^2\mathbf{I}_n$, where σ^2 is unknown positive finite parameter and $\mathbf{I}_n$ the $n \times n$ identity matrix. $E(.)$ denotes the mathematical expectation.

2.1. Unceration Prior Information

We are primarily interested in the estimation of $\boldsymbol{\beta}$ when $\boldsymbol{\beta}$ is suspected to lie in the subspace defined by

$$\mathbf{H}\boldsymbol{\beta} = \mathbf{h}, \tag{2.2}$$

where $\mathbf{H}$ is a given $q \times k$ matrix of rank $q \le k$ and $\mathbf{h}$ is a given $q \times 1$ vector of constants. This situation is a general case where the classical sub-hypothesis problem is included. Noting that a smaller rank implies inconsistency or linear dependence of the constraints that are part of (2.2).

2.2. A Large Sample Test Statistic

The likelihood Ratio (LR) test-statistic for the null hypothesis $H_o : \mathbf{H}\boldsymbol{\beta} = \mathbf{h}$ is asymptotically valid for nonnormal disturbances. Thus, we consider

$$D_n = \frac{(\mathbf{H}\hat{\boldsymbol{\beta}}^U - \mathbf{h})'[\mathbf{H}(\mathbf{X}_n'\mathbf{X}_n)^{-1}\mathbf{H}']^{-1}(\mathbf{H}\hat{\boldsymbol{\beta}}^U - \mathbf{h})}{S^2}, \tag{2.3}$$

where

$$S^2 = (\mathbf{Y}_n - \mathbf{X}_n\hat{\boldsymbol{\beta}}^U)'(\mathbf{Y}_n - \mathbf{X}_n\hat{\boldsymbol{\beta}}^U)/n - k$$

is an unbiased estimator of σ^2 and $\hat{\boldsymbol{\beta}}^U$ is the least squares estimator of $\boldsymbol{\beta}$ defined in (2.4). We have following useful results.

Theorem 2.1 Assume the elements of ϵ are independently and identically distributed with zero mean and finite variance σ^2. If the elements of X are uniformly bounded and if $\lim_{n\to\infty}(\mathbf{X}_n'\mathbf{X}_n/n) = \mathbf{Q}$ where $\mathbf{Q}$ is finite and nonsingular matrix, then

$$n^{\frac{1}{2}}(\hat{\boldsymbol{\beta}}^U - \boldsymbol{\beta}) \xrightarrow{\mathcal{L}} N(0, \sigma^2\mathbf{Q}^{-1}),$$

where $\xrightarrow{\mathcal{L}}$ means convergence in distribution.

Proof See Theil (1971, pp. 380-381).

Theorem 2.2 Assume the elements of ϵ are independent and identically distributed with zero mean and finite variance σ^2 and if $lim_{n\to\infty}(\mathbf{X}_n'\mathbf{X}_n) = \mathbf{Q}$, where $\mathbf{Q}$ is finite and nonsingular matrix. Then under the null hypothesis, $\mathbf{H}\boldsymbol{\beta} = \mathbf{h}$, the test statistic D_n follows a central chi-square distribution with q degrees of freedom.

Proof See Schmidt (1976, pp. 63-64).

2.3. Estimation Strategies

In this sub-section various estimation methods for the parameter vector $\boldsymbol{\beta}$ are proposed.

Strategy 1. Unrestricted Estimator

Based on sample information only, the *ordinary least squares estimator (OLSE)* of $\boldsymbol{\beta}$ for the model (2.1) is

$$\hat{\boldsymbol{\beta}}^U = (\mathbf{X}_n'\mathbf{X}_n)^{-1}\mathbf{X}_n'\mathbf{Y}_n = \mathbf{Q}_n^{-1}\mathbf{X}_n'\mathbf{Y}_n, \quad \text{where} \quad \mathbf{Q}_n = \mathbf{X}_n'\mathbf{X}_n. \tag{2.4}$$

Since, OLSE is unbiased and unrelated to the prior information, such an estimator is generally called a *unrestricted estimator (UE).*

Strategy 2. Restricted Estimator

The prior information given in (2.2) may be explicitly incorporated into the estimation process by modifying the parameter space. In this case, the new (restricted) parameter space is a subspace of the original one (reduced in dimension). Generally speaking, the reduction in dimensionality provides efficient parameter estimates. However, in the case of incorrect restriction opposite conclusions holds.

Under the restriction in (2.2), the *restricted least square estimator (RLSE)* or simply restricted estimator (RE) of $\boldsymbol{\beta}$ is given by

$$\hat{\boldsymbol{\beta}}^R = \hat{\boldsymbol{\beta}}^U - \mathbf{Q}_n^{-1}\mathbf{H}'(\mathbf{H}\mathbf{Q}_n^{-1}\mathbf{H}')^{-1}(\mathbf{H}\hat{\boldsymbol{\beta}}^U - \mathbf{h}). \tag{2.5}$$

If the restrictions are assumed to be correct, then $\hat{\boldsymbol{\beta}}^R$ is an unbiased estimator of $\boldsymbol{\beta}$ and superior to $\hat{\boldsymbol{\beta}}^U$. However, this may not always be the case.

Strategy 3. Shrinkage Restricted Estimator

It is reasonable to shrink $\hat{\boldsymbol{\beta}}^U$ towards $\hat{\boldsymbol{\beta}}^R$ (Thompson, 1968). Thus, a *shrinkage restricted estimator (SRE)* of $\boldsymbol{\beta}$ is defined by

$$\hat{\boldsymbol{\beta}}^{SR} = \hat{\boldsymbol{\beta}}^U - (1-\pi)(\hat{\boldsymbol{\beta}}^U - \hat{\boldsymbol{\beta}}^R),$$

where π is a coefficient reflecting degree of distrust in the prior information. Note, that $\hat{\beta}^{SR}$ is a convex combination of $\hat{\beta}^{U}$ and $\hat{\beta}^{R}$ via fixed value of π.

Strategy 4. Standard Preliminary Test Estimator

The usual *preliminary test estimator (PTE)* of β denoted by $\hat{\beta}^{P} = (\beta_1^P, \cdots \beta_k^P)$ is obtained by replacing π by $I(D_n \leq d_{n,\alpha})$ in $\hat{\beta}^{SR}$ to have a random weight. Thus,

$$\hat{\beta}^{P} = \hat{\beta}^{U} - (\hat{\beta}^{U} - \hat{\beta}^{R})I(D_n \leq d_{n,\alpha}), \tag{2.6}$$

where $I(A)$ is the indicator function of the set A and $d_{n,\alpha}$ be the upper $100\alpha\%$ $(0 < \alpha < 1)$ point of the test statistic.

Strategy 5. Shrinkage Preliminary Test Estimator

The *shrinkage preliminary test estimator (SPTE)* of β denoted by $\hat{\beta}^{SP}$ is

$$\hat{\beta}^{SP} = \hat{\beta}^{U} - (1-\pi)(\hat{\beta}^{U} - \hat{\beta}^{R})I(D_n \leq d_{n,\alpha}). \tag{2.7}$$

Strategy 6. Standard Shrinkage Estimator

The usual *shrinkage estimator (SE)* of β is defined by

$$\hat{\beta}^{S} = \hat{\beta}^{R} + \{1 - cD_n^{-1}\}(\hat{\beta}^{U} - \hat{\beta}^{R}), \quad q \geq 3, \tag{2.8}$$

where c is the shrinkage constant chosen in an interval in such a way that $\hat{\beta}^{S}$ dominates $\hat{\beta}^{U}$

Strategy 7. Improved Preliminary Test Estimator

It may be noted that $\hat{\beta}^{P}$ may also be uniformly improved by inserting Stein-type estimator instead of $\hat{\beta}^{U}$ in $\hat{\beta}^{P}$ (Sclove et al., 1972). We call it *improved preliminary test estimator (IPTE)* and is given by the following relation

$$\hat{\beta}^{P+} = \hat{\beta}^{P} - cD_n^{-1}I(D_n > d_{n,\alpha})(\hat{\beta}^{U} - \hat{\beta}^{R}), \quad q \geq 3. \tag{2.9}$$

Strategy 8. Positive-part Shrinkage Estimator

We truncate the SE in relation 2.8 with its positive-part to obtain a *positive-part shrinkage estimator (PSE)*

$$\hat{\beta}^{S+} = \hat{\beta}^{R} + \{1 - cD_n^{-1}\}^{+}(\hat{\beta}^{U} - \hat{\beta}^{R}), \quad q \geq 3, \tag{2.10}$$

where we define the notation $z^{+} = max(o, z)$. We will see that this will control the over shrinking problem inherent in $\hat{\beta}^{S}$.

The normal theory of $\hat{\beta}^{P}$ and $\hat{\beta}^{S}$ was considered by Saleh and Han (1990) and Ali and Saleh (1991). Ghosh et al. (1989) provide empirical Bayes solution to the problem. Ahmed and Ullah (1996) investigated the properties of $\hat{\beta}^{S+}$ and $\hat{\beta}^{P+}$ along with the other estimators assuming normal error. The asymptotic properties of $\hat{\beta}^P$ and $\hat{\beta}^S$ along with $\hat{\beta}^U$ and $\hat{\beta}^R$ were considered by Saleh and Sen (1989). However, the properties $\hat{\beta}^{S+}$, $\hat{\beta}^{P+}$, $\hat{\beta}^{SP}$ and $\hat{\beta}^{SR}$ are not available for the nonnormal case.

In this paper we shall attempt to provide a comprehensive study of the problem.

2.3. Asymptotic Distributional Results

We note that, as the test statistic D_n is consistent against fixed $\boldsymbol{\beta}$ such that $\mathbf{H}\boldsymbol{\beta} \neq \mathbf{h}$, the PTE, SE and proposed PSE and IPTE will be asymptotically equivalent to the $\hat{\beta}^U$ for the fixed alternative (up to the order $O(n^{-\frac{1}{2}})$) in probability. Hence, in the large sample situation there is not much to investigate. This suggest that in the asymptotic setup, we need not to consider a fixed alternative, i.e., $\mathbf{H}\boldsymbol{\beta} \neq \mathbf{h}$. To obtain the interesting and meaningful results, we shall, therefore, restrict ourselves to local alternatives. In this case these estimators are not asymptotically equivalent (up to the order $O(n^{-\frac{1}{2}})$) in probability. Specifically, we consider a sequence $\{K_n\}$ of local alternatives defined by

$$K_n : \mathbf{H}\boldsymbol{\beta} = \mathbf{h} + \frac{\boldsymbol{\delta}}{n^{\frac{1}{2}}}, \tag{2.11}$$

where $\boldsymbol{\delta} = (\delta_1, \cdots, \delta_q)' \in \Re^q$ a real fixed vector. Note that $\boldsymbol{\delta} = \mathbf{0}$ implies $\mathbf{H}\boldsymbol{\beta} = \mathbf{h}$, so (1.1) is a particular case of $\{K_n\}$. We shall compare the relative merits of the estimators under local alternatives. Towards this end, we will first briefly develop the concepts of loss, quadratic risk, and asymptotic distributional quadratic risk (ADQR) functions.

We are mainly interested in estimating the unknown parameter vector $\boldsymbol{\beta}$ by means of an estimator $\boldsymbol{\beta}_n^*$. A loss function $L(\boldsymbol{\beta}_n^*, \boldsymbol{\beta})$ will reflect the loss incurred by making wrong decision about $\boldsymbol{\beta}$ using the estimator $\boldsymbol{\beta}_n^*$. This measure will be positively related to the difference $\boldsymbol{\beta}_n^* - \boldsymbol{\beta}$. Thus, we confine ourselves to loss function of form

$$L(\boldsymbol{\beta}_n^*, \boldsymbol{\beta}; \mathbf{W}) = n(\boldsymbol{\beta}_n^* - \boldsymbol{\beta})'\mathbf{W}(\boldsymbol{\beta}_n^* - \boldsymbol{\beta}),$$

where $\mathbf{W}$ is positive semidefnite weighting matrix. Such functions are generally called *weighted quadratic loss functions.* When $\mathbf{W} = \mathbf{I}$, it is simply called *squared error loss.* Then, the expected loss functions

$$E[L(\boldsymbol{\beta}_n^*, \boldsymbol{\beta}; \mathbf{W})] = R(\boldsymbol{\beta}_n^*, \boldsymbol{\beta}; \mathbf{W}) \equiv R(\boldsymbol{\beta}_n^*, \boldsymbol{\beta}) \equiv R(\boldsymbol{\beta}_n^*),$$

are called *risk functions.* The risk function can be rewritten as

$$\begin{aligned} R(\boldsymbol{\beta}_n^*, \boldsymbol{\beta}; \mathbf{W}) &= nE\{(\boldsymbol{\beta}_n^* - \boldsymbol{\beta})'\mathbf{W}(\boldsymbol{\beta}_n^* - \boldsymbol{\beta})\} \\ &= n\,trace\,[\mathbf{W}\{E(\boldsymbol{\beta}_n^* - \boldsymbol{\beta})(\boldsymbol{\beta}_n^* - \boldsymbol{\beta})'\}] \\ &= trace(\mathbf{W}\boldsymbol{\Gamma}^*), \end{aligned} \tag{2.12}$$

where $\boldsymbol{\Gamma}^*$ is the covariance matrix of $\boldsymbol{\beta}_n^*$.

One way of measuring the performance of the estimators is to compare the risk of two estimators with a suitable matrix $\mathbf{W}$. The estimator with the smaller risk for $\boldsymbol{\beta}$ is usually preferred. Further, $\boldsymbol{\beta}_n^*$ will be termed an *inadmissible estimator* of $\boldsymbol{\beta}$ if there exists an alternative estimator $\boldsymbol{\beta}_n^o$ such that

$$\mathcal{R}(\boldsymbol{\beta}_n^o, \boldsymbol{\beta}) \leq \mathcal{R}(\boldsymbol{\beta}_n^*, \boldsymbol{\beta}) \quad \text{for all} \quad (\boldsymbol{\beta}, \mathbf{W}), \tag{2.13}$$

with strict inequality for some $\boldsymbol{\beta}$. We also say that $\boldsymbol{\beta}_n^o$ dominates $\boldsymbol{\beta}_n^*$. If, instead of (2.13) holding for every n, we have

$$\lim_{n\to\infty} \mathcal{R}(\boldsymbol{\beta}_n^o, \boldsymbol{\beta}) \leq \lim_{n\to\infty} \mathcal{R}(\boldsymbol{\beta}_n^*, \boldsymbol{\beta}) \quad \text{for all} \quad \boldsymbol{\beta}, \tag{2.14}$$

with strict inequality for some $\boldsymbol{\beta}$, then $\boldsymbol{\beta}^*$ is termed an *asymptotically inadmissible estimator* of $\boldsymbol{\beta}$. However, the expression in (2.14) may be difficult to obtain. Hence we consider the *asymptotic distributional quadratic risk (ADQR)* for a sequence $\{K_n\}$ of local alternatives defined in relation (2.11). This is based on the notion of asymptotic relative efficiency which depends on the asymptotic distribution of a sample quantity of interest. Assume that under local alternatives, $n^{\frac{1}{2}}(\boldsymbol{\beta}_n^* - \boldsymbol{\beta})$ has a limiting distribution $\{n^{\frac{1}{2}}(\boldsymbol{\beta}_n^* - \boldsymbol{\beta})\}$ given by

$$F(\mathbf{y}) = \lim_{n\to\infty} P\{\sqrt{n}(\boldsymbol{\beta}_n^* - \boldsymbol{\beta}) \leq \mathbf{y}\}, \tag{2.15}$$

which is called the *asymptotic distribution function (ADF)* of $\boldsymbol{\beta}_n^*$. Further, let

$$\boldsymbol{\Gamma} = \int\int \cdots \int \mathbf{y}\mathbf{y}' dF(\mathbf{y}), \tag{2.16}$$

be the dispersion matrix which is obtained from ADF, the ADQR may be defined as

$$R(\boldsymbol{\beta}_n^*; \boldsymbol{\beta}) = \text{trace}(\mathbf{W}\boldsymbol{\Gamma}). \tag{2.17}$$

An estimator $\boldsymbol{\beta}_n^*$ is said to dominate an estimator $\boldsymbol{\beta}_n^o$ asymptotically if, $R(\boldsymbol{\beta}_n^*; \boldsymbol{\beta}) \leq R(\boldsymbol{\beta}_n^o; \boldsymbol{\beta})$. If, in addition, $R(\boldsymbol{\beta}_n^*; \boldsymbol{\beta}) < R(\boldsymbol{\beta}_n^o; \boldsymbol{\beta})$ for at least some $(\boldsymbol{\beta}, \mathbf{W})$, then $\boldsymbol{\beta}_n^*$ strictly dominates $\boldsymbol{\beta}_n^o$. The concept of ADQR is also related to asymptotic distribution admissibility. We may be able to compute asymptotic risk by replacing $\boldsymbol{\Gamma}$ with the limit of actual dispersion matrix of $n^{\frac{1}{2}}(\boldsymbol{\beta}_n^* - \boldsymbol{\beta})$ in ADQR function. This may require extra regularity conditions to suit the problem at hand. This point has been explained in various other contexts by Sen (1984), Saleh and Sen (1985) and others. We shall study the ADQR results for the various estimators proposed in the pervious section and the remaining discussion follows.

Under local alternatives $\{K_n\}$ we have the following theorem which facilitates the computation of the asymptotic bias and ADQR of the various estimators.

Theorem 2.3 Under $\{K_n\}$ and the usual regularity conditions and as n increases we have following:

(a) $n^{\frac{1}{2}}(\hat{\boldsymbol{\beta}}^U - \boldsymbol{\beta}) \xrightarrow{\mathcal{L}} N(\boldsymbol{\delta}, \sigma^2\mathbf{Q}^{-1})$,

(b) D_n follows a non-central chi-squared distribution with q degrees of freedom and non-centrality parameter $\Delta = \frac{\boldsymbol{\delta}'\mathbf{B}^{-1}\boldsymbol{\delta}}{\sigma^2}$, where $\mathbf{B} = \mathbf{H}\mathbf{Q}^{-1}\mathbf{H}'$.

Using Theorem 2.3 and applying results from Judge and Bock (1978) given in Appendix B, we shall present the asymptotic distribution bias (ADB) and ADQR in the sub-section that follows.

2.3.1. Asymptotic Distributional Bias

First we present the expressions for *asymptotic distributional bias (ADB)* of the proposed estimators. The ADB of an estimator $\boldsymbol{\beta}_n^*$ is defined as

$$ADB(\boldsymbol{\beta}_n^*) = \lim_{n\to\infty} E\{n^{\frac{1}{2}}(\boldsymbol{\beta}_n^* - \boldsymbol{\beta})\}.$$

Theorem 2.4. Using the above definition of the ADB, under $\{K_n\}$ in (2.11) and the assumed regularity conditions, as $n \to \infty$,

$$ADB(\hat{\beta}^U) = \mathbf{0}, \tag{2.18a}$$

$$ADB(\hat{\beta}^R) = -\mathbf{Q}^{-1}\mathbf{H}\mathbf{B}^{-1}\boldsymbol{\delta}, \tag{2.18b}$$

$$ADB(\hat{\beta}^{SR}) = -(1-\pi)\mathbf{Q}^{-1}\mathbf{H}\mathbf{B}^{-1}\boldsymbol{\delta}, \tag{2.18c}$$

$$ADB(\hat{\beta}^P) = -\mathbf{Q}^{-1}\mathbf{H}\mathbf{B}^{-1}\boldsymbol{\delta} H_{q+2}(\chi^2_{q,\alpha};\Delta), \tag{2.18d}$$

$$ADB(\hat{\beta}^{SP}) = -(1-\pi)\mathbf{Q}^{-1}\mathbf{H}\mathbf{B}^{-1}\boldsymbol{\delta} H_{q+2}(\chi^2_{q,\alpha};\Delta), \tag{2.18e}$$

$$ADB(\hat{\beta}^S) = -(q-2)\mathbf{Q}^{-1}\mathbf{H}\mathbf{B}^{-1}\boldsymbol{\delta} E(\chi^{-2}_{q+2}(\Delta)), \tag{2.18f}$$

$$\begin{aligned} ADB(\hat{\beta}^{S+}) = & -\mathbf{Q}^{-1}\mathbf{H}\mathbf{B}^{-1}\boldsymbol{\delta}\big[H_{q+2}(q-2;\Delta) + (q-2)E\{\chi^{-2}_{q+2}(\Delta) + \\ & E\{\chi^{-2}_{q+2}(\Delta) I(\chi^2_{q+2}(\Delta)) > q-2)\}\big], \end{aligned} \tag{2.18g}$$

$$\begin{aligned} ADB(\hat{\beta}^{P+}) = & -\mathbf{Q}^{-1}\mathbf{H}\mathbf{B}^{-1}\boldsymbol{\delta}\big[H_{q+2}(\chi^2_{q,\alpha};\Delta) + (q-2)E\{\chi^{-2}_{q+2}(\Delta) + \\ & E\{\chi^2_{q+2};\Delta) I(\chi^2_{q+2}(\Delta)) > \chi^2_{q,\alpha})\}\big], \end{aligned} \tag{2.18h}$$

Since the bias expressions of all the estimators are not in the scalar form, we therefore take the recourse by converting them in to the quadratic form. Thus, let us define the *quadratic bias (QB)* of an estimator β^* of β by

$$QB(\beta^*) = [\mathbf{ADB}(\beta^*)]'\mathbf{Q}[\mathbf{ADB}(\beta^*)]$$

Corollary 2.1 By using definition, the quadratic bias of the various estimators is given as follows:

$$QB(\hat{\beta}^U) = 0, \tag{2.19a}$$

$$QB(\hat{\beta}^R) = \Delta, \tag{2.19b}$$

$$QB(\hat{\beta}^{SR}) = (1-\pi)^2\Delta, \tag{2.19c}$$

$$QB(\hat{\beta}^P) = \Delta\{H_{q+2}(\chi^2_{q,\alpha};\Delta)\}^2, \tag{2.19d}$$

$$QB(\hat{\beta}^{SP}) = (1-\pi)^2\Delta\{H_{q+2}(\chi^2_{q,\alpha};\Delta)\}^2, \tag{2.19e}$$

$$QB(\hat{\beta}^S) = (q-2)^2\Delta\{E(\chi^{-2}_{q+2}(\Delta))\}^2, \tag{2.19f}$$

$$\begin{aligned} QB(\hat{\beta}^{S+}) = & \Delta\big[H_{q+2}(q-2;\Delta) + (q-2)E\{\chi^{-2}_{q+2}(\Delta) + \\ & E\{\chi^{-2}_{q+2}(\Delta) I(\chi^2_{q+2}(\Delta)) > q-2)\}\big]^2, \end{aligned} \tag{2.19g}$$

$$\begin{aligned} QB(\hat{\beta}^{P+}) = & \Delta\big[H_{q+2}(\chi^2_{q,\alpha};\Delta) + (q-2)E\{\chi^{-2}_{q+2}(\Delta) + \\ & E\{\chi^2_{q+2};\Delta) I(\chi^2_{q+2}(\Delta)) > \chi^2_{q,\alpha})\}\big]^2. \end{aligned} \tag{2.19h}$$

Undoubtly, only $\hat{\boldsymbol{\beta}}^U$ is a unbiased vector estimator for the parameter vector $\boldsymbol{\beta}$. However, if the nonsample information is correct, then $\Delta = 0$ and all the remaining estimators are also unbiased. For $\Delta > 0$, the quadratic bias of $\hat{\boldsymbol{\beta}}^R$ is a function of Δ and is unbounded in Δ which goes to ∞ as Δ tends to ∞. On the other hand, quadratic bias function of all the remaining estimators are bounded in Δ.

The quadratic bias of $\hat{\boldsymbol{\beta}}^{SR}$ is a function of Δ and π. Unbiasedness is achieved when $\Delta = 0$ or when $\pi = 0$, in other words, the restriction is correctly specified or is ignored. When the hypothesis error grows then the bias of $\hat{\boldsymbol{\beta}}^{SR}$ and $\hat{\boldsymbol{\beta}}^R$ increases without a bound. However, $Bias(\hat{\boldsymbol{\beta}}^{SR}) = \pi Bias(\hat{\boldsymbol{\beta}}^R), 0 < \pi < 1$, so one may consider the shrinking technique as a bias reduction technique offered by $\hat{\boldsymbol{\beta}}^{SR}$.

As $\Delta \to \infty$ then $H_{q+2}(\chi^2_{q,\alpha}; \Delta)$ and $H_{q+4}(\chi^2_{q,\alpha}; \Delta)$ approach zero, $QB(\hat{\boldsymbol{\beta}}^{SP}) \to 0$. Further, as $\Delta \to \infty$, $QB(\hat{\boldsymbol{\beta}}^{SP})$ increases monotonically at first, reaches a maximum and then monotonically decreases towards zero. Hence it is a bounded function in Δ. It is clear that $QB(\hat{\boldsymbol{\beta}}^{SP}) = \pi QB(\hat{\boldsymbol{\beta}}^P)$. Since $0 < \pi < 1$, $QB(\hat{\boldsymbol{\beta}}^{SP}) < QB(\hat{\boldsymbol{\beta}}^P)$. This indicates that $\hat{\boldsymbol{\beta}}^{SP}$ has an edge over the pretest estimator from the quadratic bias point of view. Thus, $\hat{\boldsymbol{\beta}}^{SP}$ can be viewed as a quadratic bias reduction technique over the usual pretest estimation. Further, as π goes to zero, the behavior of the quadratic bias of the shrinkage pretest estimator is like that of the unbiased estimator, and when $\pi \to 1$, it is like the pretest estimator.

The quadratic bias functions of both $\hat{\boldsymbol{\beta}}^{S+}$ and $\hat{\boldsymbol{\beta}}^S$ starts from 0 at $\Delta = 0$ and increases to a point and then decreases towards 0 since $\mathcal{E}\left(\chi^{-2}_{p+1}(\Delta)\right)$ is a decreasing convex function of Δ. However, the graph of the quadratic bias of $\hat{\boldsymbol{\beta}}^{S+}$ remain below the graph of the quadratic bias of $\hat{\boldsymbol{\beta}}^S$.

The quadratic bias of $\hat{\boldsymbol{\beta}}^{P+}$ is a function of Δ and α. For fixed α, as a function of Δ, this function starts from 0 increases to a point, then decreases gradually to zero. On the other hand, as a function of α (for fixed Δ) it is a decreasing function of $\alpha \in [0, 1)$ which achieves a maximum value at $\alpha = 0$ while its value is 0 at $\alpha = 1$. The value of quadratic bias of $\hat{\boldsymbol{\beta}}^P$ remains either below or equal to the value of the quadratic bias of $\hat{\boldsymbol{\beta}}^{P+}$ for all values of Δ.

2.3.2. ADQR Analysis

First, we provide the risk expressions for the the estimators in the following theorem.

Theorem 2.5. Under $\{K_n\}$ in (2.11) and the assumed regularity conditions, as $n \to \infty$ the *ADQR* of the estimators are

$$R(\hat{\boldsymbol{\beta}}^U) = \sigma^2 tr(\mathbf{Q}^{-1}\mathbf{W}) \tag{2.20a}$$

$$R(\hat{\boldsymbol{\beta}}^R) = \sigma^2 tr(\mathbf{Q}^{-1}\mathbf{W}) + \boldsymbol{\delta}'\mathbf{B}^{-1}\mathbf{A}\boldsymbol{\delta} - \sigma^2 tr(\mathbf{A}) \tag{2.20b}$$

$$R(\hat{\boldsymbol{\beta}}^{SR}) = \sigma^2 tr(\mathbf{Q}^{-1}\mathbf{W}) + (1-\pi)^2\boldsymbol{\delta}'\mathbf{B}^{-1}\mathbf{A}\boldsymbol{\delta} - (1-\pi^2)\sigma^2 tr(\mathbf{A}) \tag{2.20c}$$

$$\begin{aligned} R(\hat{\boldsymbol{\beta}}^P) =& \sigma^2 tr(\mathbf{Q}^{-1}\mathbf{W}) - \sigma^2 tr(\mathbf{A}) H_{q+2}(\chi^2_{q,\alpha}; \Delta) + \\ & \boldsymbol{\delta}'\mathbf{B}^{-1}\mathbf{A}\boldsymbol{\delta}\{2H_{q+2}(\chi^2_{p,\alpha}; \Delta) - H_{q+4}(\chi^2_{q,\alpha}; \Delta)\}, \end{aligned} \tag{2.20d}$$

$$\begin{aligned} R(\hat{\boldsymbol{\beta}}^{SP}) =& \sigma^2 tr(\mathbf{Q}^{-1}\mathbf{W}) - (1-\pi^2)\sigma^2 tr(\mathbf{A}) H_{q+2}(\chi^2_{q,\alpha};\Delta) + \\ & \boldsymbol{\delta}'\mathbf{B}^{-1}\mathbf{A}\boldsymbol{\delta}(1-\pi)\{2H_{q+2}(\chi^2_{q,\alpha};\Delta) - (1+\pi)H_{q+4}(\chi^2_{q,\alpha};\Delta)\}, \end{aligned} \tag{2.20e}$$

$$\begin{aligned} R(\hat{\boldsymbol{\beta}}^{S}) =& \sigma^2 tr(\mathbf{Q}^{-1}\mathbf{W}) + \boldsymbol{\delta}'\mathbf{B}^{-1}\mathbf{A}\boldsymbol{\delta}(q^2-4)E(\chi^{-4}_{q+4}(\Delta)) - \\ & (q-2)\sigma^2 tr(\mathbf{A})\{E(\chi^{-2}_{q+2}(\Delta)) + \Delta E(\chi^{-4}_{q+4}(\Delta))\}, \end{aligned} \tag{2.20f}$$

$$\begin{aligned} R(\hat{\boldsymbol{\beta}}^{S+}) = & R(\hat{\boldsymbol{\beta}}^{S}) + (q-2)\sigma^2 tr(\mathbf{A})\big[2E\left\{\chi^{-2}_{q+2}(\Delta)I(\chi^2_{q+2}(\Delta) \le (q-2)\right\} - \\ & (q-2)E\left\{\chi^{-4}_{q+2}(\Delta)I(\chi^2_{q+2}(\Delta) \le (q-2)\right\}\big] - \sigma^2 tr(\mathbf{A})H_{q+2}(q-2;\Delta) + \\ & \boldsymbol{\delta}'\mathbf{B}^{-1}\mathbf{A}\boldsymbol{\delta}\left\{2H_{q+2}(q-2;\Delta) - H_{q+4}(q-2;\Delta)\right\} - \\ & (q-2)\boldsymbol{\delta}'\mathbf{B}^{-1}\mathbf{A}\boldsymbol{\delta}\big[2E\{\chi^{-2}_{q+2}(\Delta))I(\chi^2_{q+2}(\Delta) \le (q-2)\} - \\ & 2E\{\chi^{-2}_{q+4}(\Delta)I(\chi^2_{q+4}(\Delta) \le (q-2)\} + (q-2)E\{\chi^{-4}_{q+4}(\Delta)I(\chi^2_{q+4}(\Delta) \le (q-2)\}\big] \end{aligned} \tag{2.20g}$$

$$\begin{aligned} R(\hat{\boldsymbol{\beta}}^{P+}) = & R(\hat{\boldsymbol{\beta}}^{S}) + (q-2)\sigma^2 tr(\mathbf{A})\big[2E\left\{\chi^{-2}_{q+2}(\Delta)I(\chi^2_{q+2}(\Delta) \le \chi^2_{q,\alpha}\right\} - \\ & (q-2)E\left\{\chi^{-4}_{q+2}(\Delta)I(\chi^2_{q+2}(\Delta) \le \chi^2_{q,\alpha}\right\}\big] - \sigma^2 tr(\mathbf{A})H_{q+2}(\chi^2_{q,\alpha};\Delta) + \\ & \boldsymbol{\delta}'\mathbf{B}^{-1}\mathbf{A}\boldsymbol{\delta}\left\{2H_{q+2}(\chi^2_{q,\alpha};\Delta) - H_{q+4}(\chi^2_{q,\alpha};\Delta)\right\} - \\ & (q-2)\boldsymbol{\delta}'\mathbf{B}^{-1}\mathbf{A}\boldsymbol{\delta}\big[2E\{\chi^{-2}_{q+2}(\Delta))I(\chi^2_{q+2}(\Delta) \le \chi^2_{q,\alpha}\} - \\ & 2E\{\chi^{-2}_{q+4}(\Delta)I(\chi^2_{q+4}(\Delta) \le \chi^2_{q,\alpha}\} + (q-2)E\{\chi^{-4}_{q+4}(\Delta)I(\chi^2_{q+4}(\Delta) \le \chi^2_{q,\alpha}\}\big], \end{aligned} \tag{2.20h}$$

where $\mathbf{A} = \mathbf{H}\mathbf{Q}^{-1}\mathbf{W}\mathbf{Q}^{-1}\mathbf{H}\mathbf{B}^{-1}$.

We now investigate the statistical properties of the various estimators using *AQDR* functions and determine their dominance characteristics.

Comparison of $\hat{\boldsymbol{\beta}}^{U}$, $\hat{\boldsymbol{\beta}}^{R}$ *and* $\hat{\boldsymbol{\beta}}^{SR}$

First, note that $\hat{\boldsymbol{\beta}}^{U}$ has a constant risk since it is unrelated to the nonsample information. The $R(\hat{\boldsymbol{\beta}}^{SR})$ is an unbounded function of $\boldsymbol{\delta}$. However, it is superior to $\hat{\boldsymbol{\beta}}^{U}$ near the null hypothesis. It is seen that

$$R(\hat{\boldsymbol{\beta}}^{SR}) \le R(\hat{\boldsymbol{\beta}}^{U}) \quad \text{if} \quad (1-\pi)^2\boldsymbol{\delta}'\mathbf{B}^{-1}\mathbf{A}\boldsymbol{\delta} \le (1-\pi^2)\sigma^2 tr(\mathbf{A}).$$

Specifically, if $\boldsymbol{\delta}$ is a null vector, that is under the null hypothesis, $\hat{\boldsymbol{\beta}}^{SR}$ is superior to $\hat{\boldsymbol{\beta}}^{U}$.

In order to provide a meaningful comparison of the various estimators, we state the following theorem .

Theorem 2.6 (Courant) If **C** and **D** are two positive semi-definite.matrices with **D** nonsingular, both of order $(m \times m)$, then

$$ch_{\min}(\mathbf{C}\mathbf{D}^{-1}) \le \frac{\mathbf{x}'\mathbf{C}\mathbf{x}}{\mathbf{x}'\mathbf{D}\mathbf{x}} \le ch_{\max}(\mathbf{C}\mathbf{D}^{-1})$$

where $ch_{\min}(\cdot)$ and $ch_{\max}(\cdot)$ mean the smallest and largest eigenvalues of $(\cdot)$ respectively and $\mathbf{x}$ is a column vector of order $(m \times 1)$. We note that the above lower and upper bounds are equal to the infimum and supremum, respectively, of the ratio $\frac{\mathbf{x}'\mathbf{C}\mathbf{x}}{\mathbf{x}'\mathbf{D}\mathbf{x}}$ for $\mathbf{x} \neq \mathbf{0}$. Also, for $\mathbf{D} = \mathbf{I}$, the ratio is known as Rayleigh quotient for matrix $\mathbf{C}$.

As a consequence of the above mentioned theorem, we have

$$\sigma^2 ch_{min}(\mathbf{A}) \leq \frac{\boldsymbol{\delta}'\mathbf{B}^{-1}\mathbf{A}\boldsymbol{\delta}}{\Delta} \leq \sigma^2 ch_{max}(\mathbf{A}).$$

Thus, $R(\hat{\boldsymbol{\beta}}^R)$ intersects with the $R(\hat{\boldsymbol{\beta}}^U)$ at

$$\Delta_{\max} = \frac{(1+\pi)tr(\mathbf{A})}{(1-\pi)ch_{max}(\mathbf{A})} \quad \text{and} \quad \Delta_{\min} = \frac{(1+\pi)tr(\mathbf{A})}{(1-\pi)ch_{min}(\mathbf{A})}$$

respectively. Thus, for $\Delta \in \left[0, \frac{(1+\pi)tr(\mathbf{A})}{(1-\pi)ch_{max}(\mathbf{A})}\right]$ $\hat{\boldsymbol{\beta}}^{SR}$ has smaller risk than that of $\hat{\boldsymbol{\beta}}^U$. Alternatively, for $\Delta \in \left(\frac{(1+\pi)tr(\mathbf{A})}{(1-\pi)ch_{min}(\mathbf{A})}, \infty\right)$ $\hat{\boldsymbol{\beta}}^U$ has smaller risk. Clearly, when Δ moves away from H_o beyond the value $\frac{(1+\pi)tr(\mathbf{A})}{(1-\pi)ch_{min}(\mathbf{A})}$, the risk of $\hat{\boldsymbol{\beta}}^{SR}$ increases and becomes unbounded. This clearly indicates that the performance of $\hat{\boldsymbol{\beta}}^{SR}$ will strongly depend on the reliability of the nonsample information. The performance of $\hat{\boldsymbol{\beta}}^U$ is always steady throughout $\Delta \in [0, \infty)$. For $\pi = 0$ we obtain the comparison of $\hat{\boldsymbol{\beta}}^U$ and $\hat{\boldsymbol{\beta}}^R$. It is easily seen that if Δ is in the interval $\left[0, \frac{tr(\mathbf{A})}{ch_{max}(\mathbf{A})}\right]$ then the risk of $\hat{\boldsymbol{\beta}}^R$ is less than the risk of $\hat{\boldsymbol{\beta}}^U$. As we observed earlier, $\hat{\boldsymbol{\beta}}^{SR}$ is superior to $\hat{\boldsymbol{\beta}}^U$ when $\left[0, \frac{(1+\pi)tr(\mathbf{A})}{(1-\pi)ch_{max}(\mathbf{A})}\right]$. Thus, the range in which $\hat{\boldsymbol{\beta}}^{SR}$ is superior to $\hat{\boldsymbol{\beta}}^U$ is wider than the range in which $\hat{\boldsymbol{\beta}}^R$ performs better than $\hat{\boldsymbol{\beta}}^U$. Thus, in this sense $\hat{\boldsymbol{\beta}}^{SR}$ has an edge over $\hat{\boldsymbol{\beta}}^U$.

Now, we compare the risk of $\hat{\boldsymbol{\beta}}^{SR}$ and $\hat{\boldsymbol{\beta}}^R$. First, under the null hypothesis, i.e., $\Delta = 0$, the risk difference between two estimators

$$R(\hat{\boldsymbol{\beta}}^{SR}) - R(\hat{\boldsymbol{\beta}}^R) = \pi^2\sigma^2 tr(\mathbf{A}) > 0.$$

Hence, we conclude that, under the null hypothesis, $\hat{\boldsymbol{\beta}}^R$ performs better than $\hat{\boldsymbol{\beta}}^{SR}$. In general, $R(\hat{\boldsymbol{\beta}}^R) \leq R(\hat{\boldsymbol{\beta}}^{SR})$ for

$$\boldsymbol{\delta}'\mathbf{B}^{-1}\mathbf{A}\boldsymbol{\delta} \leq \pi(2-\pi)^{-1}\sigma^2 tr(\mathbf{A}).$$

Thus, for

$$\Delta \in \left(\frac{\pi tr(\mathbf{A})}{(2-\pi)ch_{min}(\mathbf{A})}, \infty\right)$$

$\hat{\boldsymbol{\beta}}^{SR}$ has smaller risk than that of $\hat{\boldsymbol{\beta}}^R$. The length of dominance interval depends on the value of the π.

Remark 1. In the light of above discussions, we may conclude that none of the three estimators $\hat{\boldsymbol{\beta}}^U$, $\hat{\boldsymbol{\beta}}^R$ and $\hat{\boldsymbol{\beta}}^{SR}$ dominate the other two asymptotically. However, under H_o we have the following dominance relationship $\hat{\boldsymbol{\beta}}^R \succ \hat{\boldsymbol{\beta}}^{SR} \succ \hat{\boldsymbol{\beta}}^U$, where the notation $\succ$ stands for dominance.

Comparison of $\hat{\boldsymbol{\beta}}^U$ *with* $\hat{\boldsymbol{\beta}}^{SP}$ *and* $\hat{\boldsymbol{\beta}}^P$

First, note that $H_{q+2}(\chi^2_{q,\alpha};\Delta) \leq H_{q+4}(\chi^2_{q,\alpha};\Delta) \leq H_{q+2}(\chi^2_q(\alpha);0) = 1-\alpha$, for $\alpha \in (0,1)$ and $\Delta > 0$. The left hand side of the above relation converges to 0 as $\Delta \to \infty$. Also, as $||\boldsymbol{\delta}|| \to \infty \Rightarrow \Delta \to \infty$, then $H_{q+4}(\chi^2_{q,\alpha};\Delta)$, $\boldsymbol{\delta}'\mathbf{B}^{-1}\mathbf{A}\boldsymbol{\delta}H_{q+2}(\chi^2_{q,\alpha};\Delta)$ and $\boldsymbol{\delta}'\mathbf{B}^{-1}\mathbf{A}\boldsymbol{\delta}H_{q+4}(\chi^2_{q,\alpha};\Delta)$ approach 0, and the risk of $\hat{\boldsymbol{\beta}}^{SP}$ approaches $\sigma^2 tr(\mathbf{Q}^{-1}\mathbf{W})$, i.e., the risk of $\hat{\boldsymbol{\beta}}^{U}$. The risk of $\hat{\boldsymbol{\beta}}^{SP}$ is smaller than the risk of $\hat{\boldsymbol{\beta}}^{U}$ near the null hypothesis which keeps on increasing crosses the line $\sigma^2 tr(\mathbf{Q}^{-1}\mathbf{W})$, reaches to maximum then decreases monotonically to the risk of $\hat{\boldsymbol{\beta}}^{U}$. Hence a pretest approach controls the magnitude of the risk. In fact, $\hat{\boldsymbol{\beta}}^{SP}$ dominates $\hat{\boldsymbol{\beta}}^{U}$ if $\Delta \in [0, U^*]$ where,

$$U^* = \frac{(1+\pi)tr(\mathbf{A})H_{q+2}(\chi^2_{q,\alpha};\Delta)}{ch_{max}(\mathbf{A})\{2H_{q+2}(\chi^2_{q,\alpha};\Delta) - (1+\pi)H_{q+4}(\chi^2_{q,\alpha};\Delta)\}}.$$

There are points in the parameter space for which $\hat{\boldsymbol{\beta}}^{SP}$ is inferior to $\hat{\boldsymbol{\beta}}^{U}$ and a sufficient condition is $\Delta \in (L^*, \infty)$, where

$$L^* = \frac{(1+\pi)tr(\mathbf{A})H_{q+2}(\chi^2_{q,\alpha};\Delta)}{ch_{min}(\mathbf{A})\{2H_{q+2}(\chi^2_{q,\alpha};\Delta) - (1+\pi)H_{q+4}(\chi^2_{q,\alpha};\Delta)\}}.$$

Moreover, as α (the level of significance of pretest) tends to 1, $R(\hat{\boldsymbol{\beta}}^{SP})$ tends to $R(\hat{\boldsymbol{\beta}}^{U})$. At $\Delta = 0$, the $R(\hat{\boldsymbol{\beta}}^{SP})$ assumes the value $\sigma^2 tr(\mathbf{Q}^{-1}\mathbf{W}) - \sigma^2(1-\pi^2)tr(\mathbf{A})H_{q+2}(\chi^2_{q,\alpha};\Delta)$, then keeps on increasing crossing the line $\sigma^2 tr(\mathbf{Q}^{-1}\mathbf{W})$, reaches to maximum then decreases monotonically to the $R(\hat{\boldsymbol{\beta}}^{U})$.

For $\pi = 0$, we obtain the comparison of $\hat{\boldsymbol{\beta}}^{U}$ and $\hat{\boldsymbol{\beta}}^{P}$. Thus, $\hat{\boldsymbol{\beta}}^{P}$ performs better than $\hat{\boldsymbol{\beta}}^{U}$ whenever $\Delta \in [0, U^o]$ where,

$$U^o = \frac{tr(\mathbf{A})H_{q+2}(\chi^2_{q,\alpha};\Delta)}{ch_{max}(\mathbf{A})\{2H_{q+2}(\chi^2_{q,\alpha};\Delta) - H_{q+4}(\chi^2_{q,\alpha};\Delta)\}},$$

and for $\Delta \in (L^o, \infty)$, where

$$L^o = \frac{tr(\mathbf{A})H_{q+2}(\chi^2_{q,\alpha};\Delta)}{ch_{min}(\mathbf{A})\{2H_{q+2}(\chi^2_{q,\alpha};\Delta) - H_{q+4}(\chi^2_{q,\alpha};\Delta)\}}$$

opposite conclusion holds. Further, by comparing the U^* and U^o we find that $\hat{\boldsymbol{\beta}}^{SP}$ provides a wider range than $\hat{\boldsymbol{\beta}}^{P}$ in which it has smaller risk than $\hat{\boldsymbol{\beta}}^{U}$. This indicates the superiority of $\hat{\boldsymbol{\beta}}^{SP}$ over $\hat{\boldsymbol{\beta}}^{P}$ in sense of dominance range. We will also demonstrate later in this paper that $\hat{\boldsymbol{\beta}}^{SP}$ has a remarkable edge over $\hat{\boldsymbol{\beta}}^{P}$ with respect to the size of the pretest. This important fact was first noticed by Ahmed (1992a, 1992b).

We observe that performance of the pretest estimators, which combine sample information with nonsample information, heavily depend on the correctness of the nonsample information. The gain in the risk is substantial over classical procedure when information is correct or nearly correct. However, $\hat{\boldsymbol{\beta}}^{SP}$ and $\hat{\boldsymbol{\beta}}^{P}$ combine the information in a superior way than that of $\hat{\boldsymbol{\beta}}^{SR}$ and $\hat{\boldsymbol{\beta}}^{R}$ in the sense that their risk is a bounded function of the nonsample information.

Remark 2 None of the three estimators is inadmissible with respect to each other. However, at $\Delta = 0$, the risks of the estimators may be ordered according to the magnitude of their risk as follows:

$$\hat{\beta}^{P} \succ \hat{\beta}^{SP} \succ \hat{\beta}^{U}.$$

Comparison of $\hat{\beta}^{SR}$ *and* $\hat{\beta}^{SP}$

When the nonsample information is correct, then the risk difference $R(\hat{\beta}^{SP}) - R(\hat{\beta}^{SR}) = (1-\pi^2)\sigma^2 tr(\mathbf{A})\{1 - H_{q+2}(\chi^2_{q,\alpha}; 0)\} \geq 0$. This clearly indicates superiority of $\hat{\beta}^{SR}$ over $\hat{\beta}^{SP}$ at the null hypothesis. However, under local alternative, the risk difference indicates that $\hat{\beta}^{SR}$ will be superior to $\hat{\beta}^{SP}$ if

$$0 \leq \Delta \leq \frac{(1+\pi)\{tr(\mathbf{A}) - tr(\mathbf{A})H_{q+2}(\chi^2_{q,\alpha}; \Delta)\}}{ch_{max}(\mathbf{A})\{(1-\pi) - 2H_{q+2}(\chi^2_{q,\alpha}; \Delta) + (1+\pi)H_{q+4}(\chi^2_{q,\alpha}; \Delta)\}},$$

while opposite holds if

$$\frac{(1+\pi)\{tr(\mathbf{A}) - tr(\mathbf{A})H_{q+2}(\chi^2_{q,\alpha}; \Delta)\}}{ch_{min}(\mathbf{A})\{(1-\pi) - 2H_{q+2}(\chi^2_{q,\alpha}; \Delta) + (1+\pi)H_{q+4}(\chi^2_{q,\alpha}; \Delta)\}} < \Delta < \infty.$$

The proposed estimators $\hat{\beta}^{SR}$ and $\hat{\beta}^{P}$ both use the data and nonsample information, however, neither $\hat{\beta}^{SP}$ nor $\hat{\beta}^{SR}$ is superior with respect to each other. For $\pi = 0$, we obtain the similar risk analysis of $\hat{\beta}^{P}$ and $\hat{\beta}^{R}$.

Remark 3. Under the null hypothesis the following dominance relationships are obtained. $\hat{\beta}^{R} \succ \hat{\beta}^{SR} \succ \hat{\beta}^{SP}$ and $\hat{\beta}^{R} \succ \hat{\beta}^{P}$.

Comparison of $\hat{\beta}^{SP}$ *and* $\hat{\beta}^{P}$

We now proceed to compare the risk of the SPTE and PTE and determine the conditions under which the SPTE dominates the PTE. First, under the null hypothesis, i.e., $\Delta = 0$, $R(\hat{\beta}^{SP}) - R(\hat{\beta}^{P}) = \pi^2\sigma^2 tr(\mathbf{A})H_{q+2}(\chi^2_{q,\alpha}; 0) > 0$. Thus, under the null hypothesis $\hat{\beta}^{SP}$ dominates $\hat{\beta}^{P}$. However, the risk difference may be negligible for the smaller values of π. Alternatively, when Δ deviates from the origin then $\hat{\beta}^{SP}$ dominates $\hat{\beta}^{P}$ in the rest of the parameter space. It is seen that , $\hat{\beta}^{P}$ is superior to $\hat{\beta}^{SP}$ if

$$0 \leq \Delta \leq \frac{tr(\mathbf{A})}{ch_{max}(\mathbf{A})\pi\{2H_{q+2}(\chi^2_{q,\alpha}; \Delta) - \pi H_{q+4}(\chi^2_{q,\alpha}; \Delta)\}}.$$

Further, $R(\hat{\beta}^{SP})$ intersects with the $R(\hat{\beta}^{P})$ at

$$\Delta_{\max} = \frac{tr(\mathbf{A})}{ch_{max}(\mathbf{A})\pi\{2H_{q+2}(\chi^2_{q,\alpha}; \Delta) - \pi H_{q+4}(\chi^2_{q,\alpha}; \Delta)\}} \quad \text{and}$$

$$\Delta_{\min} = \frac{tr(\mathbf{A})}{ch_{min}(\mathbf{A})\pi\{2H_{q+2}(\chi^2_{q,\alpha}; \Delta) - \pi H_{q+4}(\chi^2_{q,\alpha}; \Delta)\}},$$

Let Δ_π be a point in the parameter space at which the risk of $\hat{\beta}^{SP}$ and $\hat{\beta}^{P}$ intersect for a given π. Then, for $\Delta \in (0, \Delta_\pi]$, the PTE performs better than the SPTE, while for $\Delta \in (\Delta_\pi, \infty)$, the SPTE dominates PTE. Further, for large values of π (close to 1), the interval $(0, \Delta_\pi]$ may be negligible. Nevertheless, the SPTE and PTE share a common asymptotic property that, as $\Delta \to \infty$, their risk converge to a common limit, i.e., to the risk of $\hat{\beta}^{U}$.

Remark 4 None of $\hat{\beta}^{SP}$ and $\hat{\beta}^{P}$ is inadmissible with respect to other. At $\Delta = 0$, $\hat{\beta}^{P} \succ \hat{\beta}^{SP}$.

Finally, by combining all the remarks we have made so far, we arrive at the following conclusion.

Conclusion 1 None of the five estimators is inadmissible with respect to any of other four. However, at $\Delta = 0$, $\hat{\beta}^{SR} \succ \hat{\beta}^{R} \succ \hat{\beta}^{P} \succ \hat{\beta}^{SP} \succ \hat{\beta}^{U}$.

Comparison of $\hat{\beta}^{S}$ and $\hat{\beta}^{U}$

Now we will compare risk performance of $\hat{\beta}^{S}$ and $\hat{\beta}^{U}$. As we mentioned earlier, domination of $\hat{\beta}^{S}$ over $\hat{\beta}^{U}$ will depend on the form of the weight matrix $\mathbf{W}$. In order to determine this important fact we use the following two identities to obtain the risk of $\hat{\beta}^{S}$ in the useful form. First,

$$E\left(\chi_{q+2}^{-2}(\Delta)\right) - E\left(\chi_{q+4}^{-2}(\Delta)\right) = 2E\left(\chi_{q+4}^{-4}(\Delta)\right),$$

$$E\left(\chi_{q+2}^{-2}(\Delta)\right) - (q-2)E\left(\chi_{q+4}^{-2}(\Delta)\right) = \Delta E\left(\chi_{q+4}^{-4}(\Delta)\right).$$

Thus,

$$R(\hat{\beta}^{S}) = \sigma^2 tr(\mathbf{Q}^{-1}\mathbf{W}) - (q-2)\sigma^2 tr(\mathbf{A})\Big[(q-2)E\left(\chi_{q+4}^{-2}(\Delta)\right) + \left\{1 - \frac{(q+2)\boldsymbol{\delta}'\mathbf{B}^{-1}\mathbf{A}\boldsymbol{\delta}}{2\Delta tr(\mathbf{A})}\right\} 2\Delta E\left(\chi_{q+4}^{-4}(\Delta)\right)\Big].$$

Thus, In order to $R(\hat{\beta}^{S}) \leq R(\hat{\beta}^{U})$, for all $\boldsymbol{\delta}$, we must have

$$\left\{1 - \frac{(q+2)\boldsymbol{\delta}'\mathbf{B}^{-1}\mathbf{A}\boldsymbol{\delta}}{2\Delta tr(\mathbf{A})}\right\} \geq 0,$$

which require a design condition on $\mathbf{W}$, i.e.,

$$\mathbf{W}^* = \{\mathbf{W} : 2tr(\mathbf{A}) \geq 2(q+2)ch_{max}(\mathbf{A})\}.$$

Thus, for $\mathbf{W} \in \mathbf{W}^*$

$$R(\hat{\beta}^{S}) \leq R(\hat{\beta}^{U}), \quad \text{for all} \quad \boldsymbol{\delta},$$

with strict inequality holds for some $\boldsymbol{\delta}$. We refer to Brown and Zidek (1980) and Brown (1983) for similar results. For arbitrary choice of $\mathbf{W}$, domination result may not hold. In such a case, one may modify the shrinkage estimator $\hat{\beta}^{S}$ as

$$\hat{\beta}^{MS} = \hat{\beta}^{R} + \{\mathbf{I} - c\lambda D_n^{-1}\mathbf{W}^{-1}\mathbf{Q}\}(\hat{\beta}^{U} - \hat{\beta}^{R}),$$

where $\lambda = ch_{\min}(\mathbf{W}^{-1}\mathbf{Q})$.

In the present investigation we assume that $\mathbf{W} \in \mathbf{W}^*$ and the remaining discussion follows. The risk of $\hat{\boldsymbol{\beta}}^S$ is smaller than the risk of $\hat{\boldsymbol{\beta}}^U$ in the entire parameter space and the upper limit is attained when $\Delta \to \infty$. It clearly indicates the asymptotic inferiority of $\hat{\boldsymbol{\beta}}^U$ under local alternatives. Evidently, the largest gain in risk is achieved near the null hypothesis. The risk of $\hat{\boldsymbol{\beta}}^S$ begins with an initial value $2q^{-1}\sigma^2 tr(\mathbf{A})$ and then increases monotonically towards the risk of $\hat{\boldsymbol{\beta}}^U$. Again, it may be noted that for the dominance of $\hat{\boldsymbol{\beta}}^S$ over $\hat{\boldsymbol{\beta}}^U$ we need $q > 2$.

Comparison of $\hat{\boldsymbol{\beta}}^S$, $\hat{\boldsymbol{\beta}}^R$ and $\hat{\boldsymbol{\beta}}^{SR}$

We now wish to compare $\hat{\boldsymbol{\beta}}^S$ and $\hat{\boldsymbol{\beta}}^{SR}$, under H_o

$$R(\hat{\boldsymbol{\beta}}^S) - R(\hat{\boldsymbol{\beta}}^{SR}) = (1-\pi^2)tr(\mathbf{A}) - q^{-1}(q-2)tr(\mathbf{A}) \geq 0.$$

Thus, the risk of $\hat{\boldsymbol{\beta}}^{SR}$ is substantially smaller than the risk of $\hat{\boldsymbol{\beta}}^S$ when the null hypothesis is true. As Δ increases $E(\chi^{-4}_{q+2}(\Delta))$ decreases and the opposite conclusion holds. For $\Delta > 0$, $R(\hat{\boldsymbol{\beta}}^{SR}) \leq R(\hat{\boldsymbol{\beta}}^S)$ if

$$0 \leq \Delta \leq \frac{(1-\pi^2)tr(\mathbf{A}) - (q-2)tr(\mathbf{A})\{E(\chi^{-2}_{q+2}(\Delta)) + \Delta E(\chi^{-4}_{q+4}(\Delta))\}}{ch_{max}(\mathbf{A})\{(1-\pi)^2 - (q^2-4)E(\chi^{-4}_{q+4}(\Delta)\}}.$$

Alternatively, when Δ deviates from the null hypothesis and take values beyond the above dominance interval then $\hat{\boldsymbol{\beta}}^S$ dominates $\hat{\boldsymbol{\beta}}^{SR}$ in the rest of the parameter. More specifically, if

$$\frac{(1-\pi^2)tr(\mathbf{A}) - (q-2)tr(\mathbf{A})\{E(\chi^{-2}_{q+2}(\Delta)) + \Delta E(\chi^{-4}_{q+4}(\Delta))\}}{ch_{min}(\mathbf{A})\{(1-\pi)^2 - (q^2-4)E(\chi^{-4}_{q+4}(\Delta)\}} < \Delta < \infty,$$

then $\hat{\boldsymbol{\beta}}^S$ dominates $\hat{\boldsymbol{\beta}}^{SR}$. Hence, neither $\hat{\boldsymbol{\beta}}^S$ nor is $\hat{\boldsymbol{\beta}}^{SR}$ asymptotically superior to other under local alternatives. For $\pi = 0$ we obtain risk comparison of $\hat{\boldsymbol{\beta}}^R$ versus $\hat{\boldsymbol{\beta}}^S$.

Remark 5. Under the null hypothesis: $\hat{\boldsymbol{\beta}}^R \succ \hat{\boldsymbol{\beta}}^{SR} \succ \hat{\boldsymbol{\beta}}^S$. However, under the local alternatives the dominance picture changes to $\hat{\boldsymbol{\beta}}^S \succ \hat{\boldsymbol{\beta}}^{SR} \succ \hat{\boldsymbol{\beta}}^R$.

Comparison of $\hat{\boldsymbol{\beta}}^S$, $\hat{\boldsymbol{\beta}}^P$ and $\hat{\boldsymbol{\beta}}^{SP}$

Under the null hypothesis, the risk difference between $\hat{\boldsymbol{\beta}}^S$ and $\hat{\boldsymbol{\beta}}^{SP}$ is

$$R(\hat{\boldsymbol{\beta}}^S) - R(\hat{\boldsymbol{\beta}}^{SP}) = \sigma^2 tr(\mathbf{A})\left\{(1-\pi^2)H_{q+2}(\chi^2_{q,\alpha};0) - q^{-1}(q-2)\right\}.$$

Thus, for the above relation to be non-negative we should have $H_{q+2}(\chi^2_{q,\alpha};0) > (q-2)[q(1-\pi^2)]^{-1}$, which, in turn, requires that α lies in the set

$$C_\alpha = \{\alpha : H_{q+2}(\chi^2_{q,\alpha};0) > (q-2)[q(1-\pi^2)]^{-1}\}. \tag{2.21}$$

Thus, (2.21) specifies a range of values of α for which $\hat{\boldsymbol{\beta}}^S$ dominates $\hat{\boldsymbol{\beta}}^P$. The picture changes as $\boldsymbol{\delta}$ moves away from $\mathbf{0}$. The risk of $\hat{\boldsymbol{\beta}}^S$ and $\hat{\boldsymbol{\beta}}^{SP}$ intersect at $\Delta = \Delta_\alpha$ if the condition (2.21) is satisfied; otherwise there is no intersecting point in the parameter space. If $\Delta \in [0, \Delta_\alpha)$, then $\hat{\beta}^{SP} \succ \hat{\beta}^{S}$ while for $\Delta \in (\Delta_\alpha, \infty)$, $\hat{\beta}^{S} \succ \hat{\beta}^{SP}$. If (2.21) is not satisfied then $\hat{\beta}^{S} \succ \hat{\beta}^{SP}$. Nevertheless, $\hat{\beta}^{S}$ and $\hat{\beta}^{SP}$ both enjoy a common asymptotic property, i.e., as $\Delta \to \infty$, their risks converge to a common limit, i.e., the risk of $\hat{\beta}^{U}$. In fact, the risk of $\hat{\beta}^{S}$ is always under this asymptotic value. The risk comparison of $\hat{\beta}^{P}$ and $\hat{\beta}^{S}$ is readily obtained when $\pi = 0$.

Remark 6. When H_o is true then $\hat{\beta}^{P} \succ \hat{\beta}^{SP} \succ \hat{\beta}^{S} \succ \hat{\beta}^{U}$, for a range of α. However, dominance picture changes to $\hat{\beta}^{S} \succ \hat{\beta}^{P} \succ \hat{\beta}^{SP}$, whenever (2.21) fails to hold.

Now, we investigate the properties of $\hat{\beta}^{S+}$ and compare its risk function with the risk functions of $\hat{\beta}^{S}$ and $\hat{\beta}^{U}$ accordingly.

Comparison of $\hat{\beta}^{S+}$, $\hat{\beta}^{S}$ and $\hat{\beta}^{U}$

By examining the risk function in relation (2.20g) we conclude that

$$R(\hat{\beta}^{S+}) \leq R(\hat{\beta}^{S}) \quad \text{for all the values of } \boldsymbol{\delta},$$

with strict inequality holds for some $\boldsymbol{\delta}$. The largest risk improvement is near the null hypothesis. Thus, the risk of $\hat{\beta}^{S+}$ is also smaller than the risk of $\hat{\beta}^{U}$ in the entire parameter space and the upper limit is attained when Δ approaches to ∞. It clearly indicates the asymptotic inferiority of $\hat{\beta}^{S}$ and $\hat{\beta}^{U}$ with respect to $\hat{\beta}^{S+}$ under local alternatives. The risk of $\hat{\beta}^{S+}$ increases monotonically towards $\sigma^2 tr(\mathbf{Q}^{-1}\mathbf{W})$ from below, as Δ moves away from 0. Hence, $\hat{\beta}^{S+} \succ \hat{\beta}^{S} \succ \hat{\beta}^{U}$. We observed that the shrinkage estimators combine the sample and nonsample information in superior way since these estimators perform better than the $\hat{\beta}^{U}$ regardless the correctness of the nonsample information. However, the gain in risk over $\hat{\beta}^{U}$ is substantial when the nonsample information is nearly correct. We conclude that the proposed estimator $\hat{\beta}^{S+}$ is asymptotically superior to $\hat{\beta}^{S}$ and hence to $\hat{\beta}^{U}$.

Comparison of $\hat{\beta}^{P+}$, $\hat{\beta}^{P}$ and $\hat{\beta}^{U}$

In this sub-section we investigate the comparative properties of improved preliminary test estimator with respect to $\hat{\beta}^{U}$ and $\hat{\beta}^{P}$.

The risk function of $\hat{\beta}^{P+}$ in relation (2.20h) may be rewritten in terms of risk of $\hat{\beta}^{P}$ as

$$\begin{aligned} R(\hat{\beta}^{P+}) - R(\hat{\beta}^{P}) = &-(q-2)tr(\mathbf{A})E\left[\left\{2\chi^{-2}_{q+2}(\Delta) - (q-2)\chi^{-4}_{q+2}(\Delta)\right\}\right. \\ &\left. I\left(\chi^2_{q+2}(\Delta) > \chi_{q,\alpha}\right)\right] \\ &-(q-2)\boldsymbol{\delta}'\mathbf{B}^{-1}\mathbf{A}\boldsymbol{\delta}E\left[\left\{2\chi^{-2}_{q+4}(\Delta) - (q-2)\chi^{-4}_{q+4}(\Delta)\right\}\right. \\ &\left. I\left(\chi^2_{q+4}(\Delta) > \chi_{q\alpha}\right) 2\chi^{-2}_{q+2}(\Delta) I\left(\chi^2_{q+2}(\Delta) > \chi_{q,\alpha}\right)\right]. \end{aligned} \tag{2.22}$$

The above relation reveals that $R(\hat{\beta}^{P}) - R(\hat{\beta}^{P+}) \geq 0$, for all the values of $\boldsymbol{\delta}$, with strict inequality holds for some $\boldsymbol{\delta}$. Thus, $\hat{\beta}^{P}$ is asymptotic inferior to $\hat{\beta}^{P+}$ under local alternatives. At the null hypothesis, $R(\hat{\beta}^{P+}) < R(\hat{\beta}^{P})$. Thus, $\hat{\beta}^{P+}$ strictly dominates $\hat{\beta}^{P}$ when H_o is true.

At the null hypothesis $R(\hat{\beta}^U) - R(\hat{\beta}^P) > 0$. Hence, both pretest estimators improve on $\hat{\beta}^U$ at the null hypothesis (at the expense of poor performance elsewhere in the parameter space). The magnitude of the risk gain of pretest estimators over $\hat{\beta}^U$ at the null vector depends on the size of the test, α. As α increases, the maximum risk of $\hat{\beta}^{P+}$ and $\hat{\beta}^P$ decreases.

It is important to know that if $\chi^2_{p,\alpha} \in [0, q-2]$, then $\hat{\beta}^{P+} = \hat{\beta}^{S+}$ and hence $\hat{\beta}^{P+}$ dominates $\hat{\beta}^U$. On the other hand, $\hat{\beta}^{P+}$ behaves like the usual preliminary test estimator $\hat{\beta}^P$ whenever $\chi^2_{q,\alpha} \notin [0, q-2]$ and hence may no longer be superior to $\hat{\beta}^U$ for all values of $\boldsymbol{\delta}$.

Consider the case $\chi^2_{q,\alpha} \in (q-2, \infty)$. In this setup, the risks of both $\hat{\beta}^{P+}$ and $\hat{\beta}^P$ approaches the risk of $\hat{\beta}^U$, when $||\boldsymbol{\delta}|| \to \infty$. Hence both estimators $\hat{\beta}^{P+}$ and $\hat{\beta}^P$ share a common property that, as $||\boldsymbol{\delta}|| \to \infty$, their risks converge to a common limit, but the risk of $\hat{\beta}^{P+}$ remains always below that of $\hat{\beta}^P$ and demonstrates the non-optimality of $\hat{\beta}^P$. This clearly indicates the inferiority of $\hat{\beta}^P$ to $\hat{\beta}^{P+}$ with respect to the risk functions. Furthermore, neither $\hat{\beta}^{P+}$ nor $\hat{\beta}^P$ is superior to $\hat{\beta}^U$ in the entire parameter space, since, as $\boldsymbol{\delta}$ moves away from the null vector, the value of the risk of $\hat{\beta}^{P+}$ increases to a maximum after crossing the risk of $\hat{\beta}^U$, then decreases towards it. There are some points in the parameter space where the risk function of $\hat{\beta}^{P+}$ crosses the risk function of $\hat{\beta}^U$, and hence embraces the similar kind of criticism as being absorbed by $\hat{\beta}^P$. Again, $\hat{\beta}^{P+}$ performs uniformly better than $\hat{\beta}^U$ when $\chi^2_{q,\alpha}$ takes the value outside the interval $(q-2, \infty)$.

It is important to note that the sampling properties of the $\hat{\beta}^{SP}$ and hence that of $\hat{\beta}^P$ depend, among other factors, on the size of the test chosen for the pretest which is mostly overlooked. The significance level plays an important rule in determining the nature of the estimator. Since the size of test α is in the control of statistician, hence, we face a statistical decision problem for choosing α. In the following section, a rule for the choice of the level of significance for the pretest using efficiency criterion is discussed.

We demonstrate below that the optimal significance level is smaller for the proposed shrinkage pretest estimator than that of the pretest estimator and the smaller significance level often coincides with the traditional level of significance.

2.2.3. Size of the Pretest

The $R(\hat{\beta}^{SP})$ and $R(\hat{\beta}^P)$ both depend on α, the size of the pretest, which should be determined by the user. One method to determine α is to compute the minimum guaranteed efficiency. A method to choose a critical value according to a rule was first given in Han and Bancroft (1968) and extended by Ahmed (1992a, 1992b).

First, we introduce the notion of the asymptotic relative efficiency. The *asymptotic relative efficiency (ARE)* of an estimator β^* to another estimator β° is defined by

$$ARE(\beta^* : \beta^\circ) = R(\beta^\circ, \mathbf{W})/R(\beta^*, \mathbf{W}).$$

Bear in mind that an ARE greater than 1 indicates the degree of asymptotic superiority of β^* over β°.

Using the above definition, the asymptotic relative efficiency of $\hat{\beta}^{SP}$ with respect to $\hat{\beta}^U$

is given by

$$ARE(\hat{\beta}^{SP} : \hat{\beta}^{U}) = \Big[1 - (1-\pi^2)\frac{tr(\mathbf{A})}{\sigma^2 tr(\mathbf{Q}^{-1}\mathbf{W})} H_{q+2}(\chi^2_{q,\alpha};\Delta) + \frac{\delta'\mathbf{B}^{-1}\mathbf{A}\delta}{\sigma^2 tr(\mathbf{Q}^{-1}\mathbf{W})}\{2(1-\pi)H_{q+2}(\chi^2_{q,\alpha};\Delta) - (1-\pi^2)H_{q+4}(\chi^2_{q,\alpha};\Delta)\}\Big]^{-1}. \tag{2.23}$$

Note that for $\alpha = 0$ relation (2.23) defines the ARE of $\hat{\beta}^{R}$ relative to $\hat{\beta}^{U}$ while for $\pi = 0$, it gives the ARE of $\hat{\beta}^{P}$ relative to $\hat{\beta}^{U}$.

The $ARE(\hat{\beta}^{SP} : \hat{\beta}^{U})$ satisfies the inequality

$$(1 + ARE_{\min})^{-1} \leq ARE(\hat{\beta}^{SP} : \hat{\beta}^{U}) \leq (1 + ARE_{\max})^{-1},$$

where

$$ARE_{\min} = \frac{1}{tr(\mathbf{Q}^{-1}\mathbf{W})}\Big[ch_{min}(\mathbf{A})\Delta\{2(1-\pi)H_{q+2}(\chi^2_{q,\alpha};\Delta) - (1-\pi^2)H_{q+4}(\chi^2_{q,\alpha};\Delta)\} - (1-\pi^2)tr(\mathbf{A})H_{q+2}(\chi^2_{q,\alpha};\Delta)\Big], \tag{2.24}$$

$$ARE_{\max} = \frac{1}{tr(\mathbf{Q}^{-1}\mathbf{W})}\Big[ch_{max}(\mathbf{A})\Delta\{2(1-\pi)H_{q+2}(\chi^2_{q,\alpha};\Delta) - (1-\pi^2)H_{q+4}(\chi^2_{q,\alpha};\Delta)\} - (1-\pi^2)tr(\mathbf{A})H_{q+2}(\chi^2_{q,\alpha};\Delta)\Big]. \tag{2.25}$$

Further, when the null hypothesis is true then

$$ARE(\hat{\beta}^{SP} : \hat{\beta}^{U}) = \left\{1 - \frac{(1-\pi^2)tr(\mathbf{A})}{tr(\mathbf{Q}^{-1}\mathbf{W})}\right\}^{-1} > 1.$$

Thus, the maximum value of $ARE(\hat{\beta}^{SP} : \hat{\beta}^{U})$ occurs at $\Delta = 0$. This maximum efficiency is decreasing function of α for fixed π and of π for fixed α.

However, for nonnull situation the efficiency picture is somewhat different. Noting that, for fixed values of α and π, $ARE_{\max}$ is a monotone decreasing function of Δ, where α and π are held constant; it crosses the line $ARE_{\max} = 1$ at L^*, where

$$L^* = \frac{(1+\pi)tr(\mathbf{A})H_{q+2}(\chi^2_{q,\alpha};\Delta)}{ch_{min}(\mathbf{A})\{2H_{q+2}(\chi^2_{q,\alpha};\Delta) - (1+\pi)H_{q+4}(\chi^2_{q,\alpha};\Delta)\}}.$$

then decreases and attains a minimum value E_o at a point Δ_o and then increases asymptotically to 1. The minimum efficiency is an increasing function of α. On the other hand, for any fixed value of α, the maximum value of the $ARE(\hat{\beta}^{SP} : \hat{\beta}^{U})$ denoted by E^* is a decreasing function of π, while the minimum efficiency E_o is an increasing function of π.

In order to determine the critical value for the pretest with minimum guaranteed efficiency E_o the researcher is willing to accept for fixed π, one needs to solve the equation

$$\sup_{\alpha}\left\{\inf_{\Delta} ARE_{\max}(\alpha, \pi, \Delta)\right\} \geq E_o,$$

or

$$\sup_{\alpha_1}\left\{\inf_{\Delta} ARE_{\max}(\alpha,\pi,\Delta)\right\} \geq E_o,$$

where α_1 is the required value. In the same way, by selecting $\pi = 0$, we obtain α_2 by using

$$\sup_{\alpha_2}\left\{\inf_{\Delta} ARE_{\max}(\alpha,\pi,\Delta)\right\} \geq E_o,$$

where α_2 is the desired value. Finally, by the property of $\inf_{\Delta} ARE_{\max}(\alpha,\pi,\Delta)$, we observe that $\alpha_1 < \alpha_2$. Hence, $\hat{\boldsymbol{\beta}}^{SP}$ has a remarkable edge over $\hat{\boldsymbol{\beta}}^{P}$ with respect to the size of the pretest. The use of the usual pretest estimation may be limited by the larger value of α, the level of significance. However, by employing a shrinkage technique one may control the value of size of the test. Ahmed (1992a) was first to discover this important feature of $\hat{\boldsymbol{\beta}}^{SP}$.

2.2.4. Numerical Example

In an effort to calculate the efficiency of $\hat{\boldsymbol{\beta}}^{SP}$ relative to $\hat{\boldsymbol{\beta}}^{U}$, we consider $\mathbf{W} = \sigma^{-2}\mathbf{Q}$. Hence

$$\begin{aligned} ARE(\hat{\boldsymbol{\beta}}^{SP} : \hat{\boldsymbol{\beta}}^{U}) = \Big[& 1 - (1-\pi^2)H_{q+2}(\chi^2_{q,\alpha};\Delta) + \\ & \frac{\Delta}{q}(1-\pi)\{2H_{q+2}(\chi^2_{q,\alpha};\Delta) - (1+\pi))H_{q+4}(\chi^2_{q,\alpha};\Delta)\}\Big]^{-1}. \end{aligned} \tag{2.26}$$

The maximum efficiency is achieved at $\Delta = 0$ with a value given by

$$E^* = \frac{1}{1-(1-\pi^2)H_{q+2}(\chi^2_{q,\alpha};0)} > 1.$$

In order to provide a numerical example we compute the table of maximum (E^*) and minimum (E_o) efficiencies along with corresponding value of the $\Delta = \Delta_o$ at which minimum occurred. The purpose of preparing the tables is two fold; one is to examine the performance of the estimators and the other is to determine the optimum choice of α for the pretest which provides the minimum guaranteed efficiency (E_o).

As an example, if $q = 2$ and researcher is looking for an estimator with a minimum *ARE* of at least 0.91, with $\pi = 0.6$ then from table 1 the value of α_1 is found to be 0.05. Such a choice of α_1 would yield an estimator with a maximum efficiency of 2.05 at $\Delta = 0$, with a minimum guaranteed efficiency of 0.91. If the experimenter selects $\pi = 0$, then from table 1 the size of the pretest, i.e., the value of α_2 will be approximately 0.30. Hence, $\hat{\boldsymbol{\beta}}^{SP}$ outperforms $\hat{\boldsymbol{\beta}}^{P}$ with respect to the size of the preliminary test. Not only that, the maximum efficiency drops from 2.05 to 1.51. We conclude this section with the following remark.

Remark 6: The proposed $\hat{\boldsymbol{\beta}}^{SP}$ has an edge over $\hat{\boldsymbol{\beta}}^{P}$ with respect to smaller level of significance as well as wider range of dominance over $\hat{\boldsymbol{\beta}}^{U}$. Hence, $\hat{\boldsymbol{\beta}}^{SP}$ is more efficient than $\hat{\boldsymbol{\beta}}^{P}$.

Table 1: *Maximum and Minimum Efficiencies.*

α/π	0	.1	.2	.3	.4	.5	.6	.7	.8	♣
	17.84	15.27	10.66	7.09	4.83	3.42	2.53	1.93	1.51	$E*$
.01	0.51	0.56	0.62	0.67	0.73	0.79	0.84	0.89	0.94	E_o
	8.8	9.2	9.6	9.6	10.4	10.8	11.2	12.0	12.0	Δ_o
	5.00	4.81	4.31	3.68	3.05	2.50	2.05	1.69	1.40	$E*$
.05	0.66	0.71	0.75	0.79	0.83	0.87	0.91	0.94	0.96	E_o
	6.8	7.2	7.2	7.6	8.0	8.4	8.4	8.8	9.2	Δ_o
	3.03	2.97	2.80	2.56	2.29	2.01	1.75	1.52	1.32	$E*$
.10	0.75	0.78	0.82	0.85	0.88	0.91	0.93	0.96	0.97	E_o
	6.0	6.0	6.4	6.8	6.8	7.2	7.6	8.0	8.0	Δ_o
	2.30	2.27	2.19	2.06	1.90	1.74	1.57	1.40	1.26	$E*$
.15	0.80	0.83	0.86	0.89	0.91	0.93	0.95	0.97	0.98	E_o
	5.6	5.6	6.2	6.4	6.4	6.8	7.2	7.2	7.2	Δ_o
	1.92	1.90	1.85	1.77	1.67	1.56	1.44	1.32	1.21	$E*$
.20	0.84	0.87	0.89	0.91	0.93	0.95	0.96	0.97	0.98	E_o
	5.2	5.2	5.6	6.0	6.0	6.4	6.4	6.4	6.8	Δ_o
	1.68	1.66	1.63	1.58	1.51	1.43	1.35	1.26	1.17	$E*$
.25	0.87	0.89	0.91	0.93	0.94	0.96	0.97	0.98	0.99	E_o
	4.8	5.2	5.2	5.6	6.0	6.0	6.0	6.0	6.0	Δ_o
	1.51	1.50	1.48	1.45	1.40	1.34	1.28	1.21	1.14	$E*$
.30	0.90	0.91	0.93	0.94	0.95	0.96	0.97	0.98	0.99	E_o
	4.8	4.8	5.2	5.2	5.6	5.6	5.6	6.0	6.0	Δ_o
	1.31	1.30	1.29	1.27	1.24	1.21	1.18	1.13	1.09	$E*$
.40	0.93	0.94	0.95	0.96	0.97	0.98	0.98	0.99	0.99	E_o
	4.4	4.4	4.4	4.8	4.8	4.8	5.2	5.2	5.6	Δ_o
	1.18	1.17	1.17	1.16	1.15	1.13	1.11	1.08	1.06	$E*$
.50	0.96	0.97	0.97	0.98	0.98	0.99	0.99	0.99	0.99	E_o
	4.0	4.0	4.4	4.4	4.4	4.8	4.8	4.8	4.8	Δ_o

The computations for table has been carried out with a FORTRAN program.

2.2.5 The Prediction

First, we wish to remark here that word *prediction* may be misleading and generally it is not applicable in its real meanings. In the regression context, it is used as conditional estimation of the mean response in a regression model. In sequel, we will continue to use this word in the sense of the conditional estimation of the mean response in a regression model.

As stated earlier the risk functions of the estimators depends on $\mathbf{Q}$. Thus, the dominance pictures of the estimators were data specific. To avoid this complication, we have set $\mathbf{W} = \sigma^{-2}\mathbf{Q}$ in our numerical work. However, this choice of $\mathbf{W}$ may not be appropriate in the context of prediction, we refer to Brown (1983). Further, in order to avoid the data dependent condition in comparing the risks of the estimators, many authors have concentrated on the conditional prediction rather than the estimation $\boldsymbol{\beta}$ itself. In this case, it is equivalent to assume that $\mathbf{Q} = \mathbf{I}$. In other words we are assuming orthonormal regressors.

From model (2.1) the prediction parameter vector is given by

$$E\left(\mathbf{y}_n|\mathbf{X}_n\right) = \mathbf{X}_n\boldsymbol{\beta},$$

which for convenience of notation will be denoted by $\boldsymbol{\mu}$. Then the risk of $\hat{\boldsymbol{\mu}}^U$, unrestricted estimator of $\boldsymbol{\mu}$ may be deduced from the risk of the $\hat{\boldsymbol{\beta}}^U$ by observing

$$\begin{aligned} R(\hat{\boldsymbol{\mu}}^U, \boldsymbol{\mu}) &= E\{(\hat{\boldsymbol{\mu}}^U - \boldsymbol{\mu})'(\hat{\boldsymbol{\mu}}^U - \boldsymbol{\mu})'\} \\ &= E\{(\mathbf{X}_n\hat{\boldsymbol{\beta}}^U - \mathbf{X}_n\boldsymbol{\beta})'(\mathbf{X}_n\hat{\boldsymbol{\beta}}^U - \mathbf{X}_n\boldsymbol{\beta})\} \\ & E\{(\hat{\boldsymbol{\beta}}^U - \boldsymbol{\beta})'\mathbf{X}_n'\mathbf{X}_n(\hat{\boldsymbol{\beta}}^U - \boldsymbol{\beta})\} = \sigma^2 q. \end{aligned}$$

Now, we define various other estimators of $\boldsymbol{\mu}$ when it is suspected that $\mathbf{H}\boldsymbol{\beta} = \mathbf{h}$. The SRE, SPTE, SE, PSE, and IPTE of $\boldsymbol{\mu}$ are given below:

$$\begin{aligned} \hat{\boldsymbol{\mu}}^{SR} &= \hat{\boldsymbol{\mu}}^U + (1-\pi)(\hat{\boldsymbol{\mu}}^U - \hat{\boldsymbol{\mu}}^{SR}), \\ \hat{\boldsymbol{\mu}}^{SP} &= \hat{\boldsymbol{\mu}}^U - (1-\pi)(\hat{\boldsymbol{\mu}}^U - \hat{\boldsymbol{\mu}}^R)I(D_n \leq d_{n,\alpha}), \\ \hat{\boldsymbol{\mu}}^S &= \hat{\boldsymbol{\mu}}^R + \{1 - cD_n^{-1}\}(\hat{\boldsymbol{\mu}}^U - \hat{\boldsymbol{\mu}}^R), \quad q \geq 3, \\ \hat{\boldsymbol{\mu}}^{S+} &= \hat{\boldsymbol{\mu}}^R + \{1 - cD_n^{-1}\}^+(\hat{\boldsymbol{\mu}}^U - \hat{\boldsymbol{\mu}}^R), \quad q \geq 3 \\ \hat{\boldsymbol{\mu}}^{P+} &= \hat{\boldsymbol{\mu}}^P - cD_n^{-1}I(D_n > d_{n,\alpha})(\hat{\boldsymbol{\mu}}^U - \hat{\boldsymbol{\mu}}^R), \quad q \geq 3. \end{aligned}$$

The RE and PTE may be obtained by using $\pi = 0$ in SRE and in SPTE respectively.

3. Concluding Remarks and Outlook for Future

We have discussed various estimation strategies for combining nonsample and sample information and presented a large sample perspective in quite generality. Various estimators of the parameter of interest are proposed by applying pretest and Stein-type estimation. It is concluded that positive-part shrinkage estimator is more efficient than the usual shrinkage estimator and they both perform well relative to standard unrestricted estimator of the parameter vector in the entire parameter space. In contrast, the performance of the estimators based on pretest estimation and restricted estimation heavily depends on the quality of the nonsample information. However, the risk of restricted estimator is unbounded when the parameter moves far from the subspace of the restriction. On the other hand, pretest estimators have a good control on the magnitude of the risk. Further, the shrinkage estimators has the smallest possible risk in most cases as compared to other estimators except when the nonsample information is nearly correct. In such cases, the estimators generated by restricted and pretest rule are superior. It is noted that the application of shrinkage estimators are subject to condition that $q > 2$. Therefore, we recommend the use of pretest estimators when such a constraint holds. However, when $q > 2$, from the point of robust performance, use of all the estimators may be advocated leaning towards shrinkage estimators.

Professor Efron in *RSS News of January, 1995* writes:

"The empirical Bayes/James-Stein category was the entry in my list least affected by computer developments. It is a ripe for a computer-intensive treatment that brings the substantial benefits of James-Stein estimation to bear on complicated, realistic problems. A side

benefit may be at least a partial reconciliation between frequentist and Bayesian perspective as they apply to statistical practice."

It may be worth mentioning that this is one of the two areas Professor Efron predicted for the early 21st century. Finally, the shrinkage/empirical Bayes and likelihood-based methods continue to play vital roles in statistical inference. These methods provide extremely useful techniques for combining data from various sources. Asymptotic theory has advanced understanding of the fundamental role of the likelihood function.

In the end, we wish to remark here that the risk analysis we have provided is based on the asymptotic results. It is also important to investigate the properties of the proposed estimators for small samples when error disturbance are nonnormal. It appears that an analytical treatment of the ***quadratic bias and risk*** performances of the proposed estimators is difficult to asses in such cases. An extensive simulation study is under way by the author to appraise the performance of the proposed estimators when sample size is small. In the process, the significance level and the power of the test of the proposed large sample test statistics will be also investigated.

ACKNOWLEDGMENTS

This work was supported by grants from the Natural Sciences and Engineering Research Council of Canada. The author is thankful to Shoal M. Khan for helpful comments that led to the improvement in the presentation of the paper.

REFERENCES

Ahmed, S.E. (1991a). To pool or not to pool: The discrete data. Statistics and Probability Letters, **11**, 233-237.

Ahmed, S.E. (1991b). Combining Poisson means. Communications in Statistics - Theory and Methods, **20**, 771-789.

Ahmed, S. E.(1992a). Shrinkage preliminary test estimation in multivariate normal distributions. Journal of Statistical Computation and Simulation **43**, 177-195.

Ahmed, S. E.(1992b). Large-sample pooling procedure for correlation. The Statistician **41**, 425-438.

Ahmed, S. E. (1993). Pooling means under uncertain prior information with applications to discrete distributions. Statistics, **24** 265-277.

Ahmed, S. E.(1994a). Improved estimation in a multivariate regression model. Journal of Computational Statistics and Data Analysis, **17**, 537-554.

Ahmed, S. E.(1994b). Pooling reliability functions. Journal of Statistical Computation and Simulation, **49**, 85-101.

Ahmed, S. E.(1994c). Improved estimation of the coefficient of variation. Journal of Applied Statistics, **21**, 565-573.

Ahmed, S. E.(1995a) A pooling methodology for coefficient of variation. Sankhya. Series B **57**, 57-75.

Ahmed, S.E. (1995b). Improved shrinkage estimation of relative potency. Biometrical Journal **37**, 627 - 638.

Ahmed, S. E. (1996a). Simultaneous estimation of survivor functions in exponential lifetime models. Submitted for publication.

Ahmed, S. E. (1996b). Simultaneous estimation of coefficients of variation. Submitted for publication.

Ahmed, S.E. and W. J. Braun (1996). Tumor growth rate pretest estimation. Proceeding of the Fifth Islamic Countries Conference on Statistical Sciences Brawijaya University, Malang Indonesia. 24-31 August 1996.

Ahmed, S.E. and A. K. Gupta (1996). Simultaneous estimation of several intraclass correlation coefficients. Submitted for publication.

Ahmed, S.E. and S. M. Khan (1991). Shrinkage estimation in pooling data for arbitrary populations. Environmetrics, **2**, 457-474.

Ahmed, S. E. and S. M. Khan (1996). Combining proportions from several randomized response model. Submitted for publication.

Ahmed S. E. and Kulperger, R. J.(1990). Asymptotic confidence intervals from preliminary test estimator. Environmetrics **1**, 295-303.

Ahmed, S. E. and V.K. Rohatgi (1996). Shrinkage estimation of proportion in randomized response model. Metrika **43**, 17- 30.

Ahmed, S. E. and A.K.Md.E. Saleh (1996). Improved nonparametric estimation of location vector in a multivariate regression model. Submitted for publication.

Ahmed, S. E. and A.K.Md.E. Saleh (1989). Pooling multivariate data. Journal of Statistical Computation and Simulation **35**, 209-226.

Ahmed, S.E. and A.K.Md.E. Saleh (1990). Estimation strategies for the intercept vector in a simple linear multivariate normal regression model. Computational Statistics and Data Analysis, **10**, 193-206.

Ahmed, S.E. and A.K.Md. E. Saleh (1993). Improved estimation for the component mean-vector. Japan Journal of Statistics, **43**, 177-195.

Ahmed, S. E. and J. S. Siddique (1995). Modifying median estimator. Pakistan Journal of Statistics, **11**, 117-122.

Ahmed, S.E. and R. J. Tomkins (1995). Estimating lognormal means using uncertain prior information. Pakistan Journal of Statistics **11**, 67-92.

Ahmed, S.E. and R. J. Tomkins (1996). On pooling means from two lognormal populations with unequal variances. Submitted for publication.

Ahmed, S.E. and B. Ullah (1996a). Improved biased estimation in an ANOVA model. Submitted for publication.

Ahmed, S.E. and B. Ullah (1996b). On some shrinkage estimators of regression coefficients under uncertain restriction in subspace. In preparation.

Ali, A. M. and A.K.Md. E. Saleh (1991). Asymptotic theory of simultaneous estimation of binomial means. Statistica Sinica, **1**, 271-294.

Ali, A. M. and A.K.Md. E. Saleh (1991). Preliminary test and empirical Bayes to shrinkage estimation of regression parameters. Journal of Japan Statistical Society, **21**, 401-416.

Anderson, T.W. (1984). An Introduction to Multivariate Statistical Analysis. New York:Wiley.

Arabatzis, A. A., T.G. Gregorie and M. R. Reynolds (1989). Conditional interval estimation of the mean following rejection of a two side test. Communications in Statistics–Theory and Methods, **18**, 4359-4373.

Bancroft, T.A. (1944). On biases in estimation due to the use of preliminary tests of significance. Annals of Mathematical Statistics, **15**, 190-204.

Bancroft, T.A and Han, C.P. (1977). Inference based on conditional specification: A note and a bibliography. International Statistical Review, **45**, 117-127.

Berger, J. O. (1985). Statistical decision theory and Bayesian analysis, second edition. Springer-Verlag: New York.

Berger, J. O., M. E. Bock, L.D. Brown., G. Casella and L. Glasser.(1977). Minimax estimation of a normal mean vector for arbitrary quadratic loss and unknown covariance matrix. Annals of Statistics, **8**, 716-761.

Brandwein, A.C. and W.E. Strawderman (1990). Stein estimation: The spherically symmetric case. Statistical Science, **5**, 356-369.

Brekson, J. (1942). Test of significance considered as evidence. Journal of American Statistical Association, **37**, 325-335.

Brown P. J. (1983). Discussion: Contribution to paper by Copas. Journal of Royal Statistical Society B, 311-354.

Brown, P.J. and J. Zidek (1980). Adaptive multivariate ridge regression. Annals of Statistics, **8**, 64-74.

Chen, J. and J.T. Hwang (1988). Improved set estimators for the coefficient of a linear model where error distribution is spherically symmetric with unknown variances. Canadian Journal of Statistics, **16**, 293-299.

Chiou, P. and C. P. Han (1994). Conditional interval estimation of the exponential scale parameter following rejection of a preliminary test. Journal of Statistical Research, **28**, 65-72.

Chiou, P. and C. P. Han (1995). Conditional interval estimation of the exponential location parameter following rejection of a pre-test. Communications in Statistics–Theory and Methods, **24**, 1481-1924.

Dasgupta, A. (1977). A general theorem in decision theory for nonnegative functional: with applications. Annals of Statistics, **17**, 1360-1374.

Efron, B. and C. N. Morris (1975). Data analysis using Stein's estimators and its generalizations. Journal of American Statistical Association, **70**, 311-319.

Efron, B. and C. N. Morris (1977). Stein's paradox in statistics. Scientific American, **236**, 119-127.

Ghosh, M, A.K.Md.E. Saleh, and P.K. Sen (1989). Empirical Bayes subset estimation in regression models. Statistics and Decisions **7**, 15-35.

Gupta, A. K., A.K.Md.E. Saleh and P.K. Sen (1989). Improved estimation in a contingency table: independence structure. Journal of American Statistical Association, **84**, 525-532.

Han, C.P. and T.A. Bancroft (1968). On pooling means when variance is unknown. Journal of American Statistical Association, **63**, 1333-1342.

Han, C.P., C.V. Rao and J. Ravichandran (1988). Inference based on conditional specification: A second bibliography. Communications in Statistics–Theory and Methods **17**, 1945-1964.

Haff, L. R. (1979). An identity for the Wishart distribution with applications. Journal of Multivariate Analysis **9**, 531- 542.

Hoffman, K. (1992). Improved estimation of distribution parameters: Stein-Type estimators. B. G. Teubner Vrlgagesellschaft: Stuttgart.

Hwang, J.T. and G. Cassela (1982). Minimax confidence set for the mean of a multivariate normal distribution. Annals of Statistics, **10**, 868-881.

Hwang, J.T. and G. Cassela (1984). Improved set estimation for a multivariate normal mean. Statistics and Decision, **1**, 3-16.

Hwang, J.T. and J. Chen (1986). Improved confidence sets for the coefficient of a linear model with spherically symmetric errors. Annals of Statistics, **14**, 444-460.

Hwang, J.T. and A. Ullah (1994). Confidence sets centered at James-Stein estimators. Journal of Econometrics, **60**, 145-156.

James W. and Stein C. (1961). Estimation with quadratic loss. Proceeding of the fourth Berkeley symposium on Mathematical statistics and Probability. University of California Press. 361-379.

Judge, G. G. and M. E. Bock (1978). The statistical implication of pre-test and Stein-rule estimators in econometrics. North-Holland:Amsterdam.

Koul, H.L. and A.K.Md.E. Saleh (1993). R-estimation of the parameters of autoregressive [AR(p)] models. Annals of Statistics, **21**, 534-551.

Kubokawa, T., T. Honda, T. Morita, and A. K. Md. E. Saleh (1993). Estimating a covariance matrix of normal distribution with unknown mean. Journal of Japan Statistics Society, **23**, 131-144.

Kulperger, R. J. and S. E. Ahmed (1992). A bootstrap theorem for a preliminary test estimator. Communications in Statistics - Theory and Methods **21**, 2071-2082.

Meeks, S.L. and D'Agostino, R. B. (1983). A note on the use of confidence limits following rejection of a null hypothesis. The American Statistician, **37**, 134-136.

Robert, C. P. (1994). The Bayesian choice: A decision-theoretic motivation. Springer-Verlag: New York.

Robins, H. (1983). Some thoughts on empirical Bayes estimation. Annals of Statistics, **11**, 713- 723.

Rukhin, A.L. (1995). Admissibility: survey of concept in progress. International Statistical Review, **63**, 95-115.

Saleh, A.K.Md.E. and C. P. Han (1990). Shrinkage estimation in regression Analysis. Estadistica, **42**, 40-63.

Saleh, A.K.Md.E. and P.K. Sen (1978). Non-parametric estimation of location parameter after a preliminary test regression. Annals of Statistics, **6**, 154-168.

Saleh, A.K.Md.E. and P.K. Sen (1984). Least square and rank order preliminary test estimation in general multivariate linear models. *Proc of ISI, Golden Jubilee Conference on Statistics: Application and New Directions (December 16-19, 1981)*, 237-253.

Saleh, A.K.Md.E. and P.K. Sen (1985a). On shrinkage M- estimator of location parameters. Communications in Statistics–Theory and Methods, **14**, 2313-2329.

Saleh, A.K.Md.E. and P.K. Sen (1985b). Preliminary predicted in general multivariate linear models. Proc. of the Pacific area Statistical Conference (edited by M. Matusita, 1982), 619-638. Amsterdam:North Holland. location parameters.

Saleh, A.K.Md.E. and P.K. Sen (1985c). Shrinkage least square estimation in a general multivariate linear model. Proc. of the 5th pannonian Symp. on Math. Statisticas (edited by J. Mogyorodi, I. Vincze, and W. Wertz1982), 307-325.

Saleh, A.K.Md.E. and P.K. Sen (1986). On shrinkage least-squares estimation in a parallelism problem. Communications in Statistics–Theory and Methods, **15**, 1451-1466.

Saleh, A.K.Md.E. and P.K. Sen (1987). Estimation of regression parameters under uncertain prior restriction in a subspace. A Festschrift in celebration of Professor M. M. Ali's 25 years at Western (edited by M.S. Haq and S.B. Provost), 171-188.

Schmidt, P. (1976). Econometrics. New York:Mercel Dekker.

Sclove, S. L., Morris, C. and R. Radhakrishnan (1972). Non-otimality of preliminary test estimators of the mean of a multivariate normal distribution. Annals of Mathematical Statistics, **43**, 1481- 1490.

Sen, P.K. (1986). On the asymptotic distributional risk shrinkage and preliminary test versions of maximum likelihood estimators. Sankhya A, **48**, 354-371.

Sen, P.K and Saleh, A.K.Md.E. (1979). Non-parametric estimation of location parameter after a preliminary test regression in a multivariate case. Journal of Multivariate Analysis **9**, 322-331.

Sen, P.K. and A.K.Md.E. Saleh (1985). On some shrinkage estimators of multivariate location. Annals of Statistics, **13**, 272-281.

Stein, C. (1956). Inadmissibility of the usual estimator for the mean of a multivariate normal distribution. Proceeding of the fourth Berkeley symposium on Mathematical Statistics and Probability. University of California Press. volume 1, 197-206.

Stigler, S. M. (1990). The 1988 Neyman Memorial Lecture: A Galtonian perspective on shrinkage estimators. Statistical Science **5**, 147-155.

Theil, H. (1971). Principles of Econometrics. New York:Wiley .

Thompson, J.C. (1968). Some shrinkage techniques for estimating the mean. Journal of American Statistical Association **63**, 113-122.

Applied Statistical Science, II
ISBN 1-56072-469-2

ESTIMATION OF PARAMETERS USING MODIFIED RANKED SET SAMPLING

Dinesh S. Bhoj
Department of Mathematical Sciences
Rutgers University

SUMMARY

A modified ranked set sampling procedure is used to derive the estimators for the location and scale parameters of some distributions. These estimators are compared with those obtained from the ranked set sampling procedure and the ordered least squares method. It is shown that the relative precisions of these estimators are substantially higher than those based on ranked set sampling procedure and ordered least squares method.

Key words: Least squares estimators; Minimum variance linear unbiased estimators; Ordered observations; Logistic distribution; Rectangular distribution; Relative precision.

1. INTRODUCTION

The concept of ranked set sampling (RSS) was first introduced by McIntyre (1952) in relation to estimating pasture yields. McIntyre's goal was to obtain an unbiased estimator for the population mean with greater efficiency than simple random sampling (SRS). He proposed RSS, which is a method of sampling that can be advantageous when quantification of all sampling units is costly but where small sets of units can be ranked according to the character under investigation by means of visual inspection or other methods not requiring actual measurements. Halls and Dell (1966) used this method in estimating forage yields, and they concluded that the procedure was more efficient than SRS to estimate population mean. Dell and Clutter (1972) and Takahasi and Wakimoto (1968) provided mathematical foundation for RSS. Dell and Clutter (1972) also showed that the estimator for the population mean based on RSS is at least as efficient as the SRS estimator with the same number of quantifications even when there are ranking errors. David and Levine (1972) considered the case where the ranking is done on the basis of a covariate instead of judgment. Under certain assumptions, they obtained a formula in which the relative precision is expressed in terms of the squared correlation coefficient between the covariate and the variate of interest. Stokes (1977) explored this model further. Stokes (1980) used RSS to obtain an asymptotically unbiased estimator for the population variance. Stokes and Sager (1988) derived a RSS estimator for a cumulative distribution function, and it was used to construct confidence bands for the function. Bohn and Wolfe (1992) constructed distribution free competitors to the Mann-Whitney estimation and testing procedures by using RSS. Recently Stokes (1995) considered both maximum likelihood and best linear unbiased estimation of location and scale parameters of a certain family of distributions. She also examined some modifications of RSS to obtain better estimators. In general, there are no closed form maximum likelihood estimators for location and scale parameters. However, the standard errors of these estimators can be computed by making use of the estimated Fisher information.

Patil, Sinha and Taillie (1993a) used the RSS method when sampling is from a finite population. They gave explicit expressions for the variance and relative precision of RSS estimator for several set sizes when the population follows a linear or quadratic trend. These authors (1993b) also studied the relative precision of RSS estimator with the regression estimator when ranking is done on the basis of an auxiliary variable. Recently Patil et al (1995) applied RSS for estimating the population mean when sampling is without replacement from a completely general finite population. They obtained explicit expressions for the variance of the RSS estimator for the mean and for its relative precision to that of random sampling without replacement. Muttlak and McDonald (1990) developed the RSS theory when the experimental units are selected with size-biased probability of selection. Yanagawa and Shirahata (1976) generalized McIntyre's (1952) RSS procedure by introducing a Selective Probability Matrix. Recently Muttlak (1995) obtained the estimators of the slope and intercept of the linear regression line by using RSS.

Recently Bhoj and Ahsanullah (1996a) derived the minimum

variance linear unbiased estimators for the parameters of the generalized geometric distribution using RSS. Bhoj (1996a) also used RSS and its suitable modification to derive the estimators of mean and standard deviation of Laplace and normal distributions. He (1996b) also obtained the best linear unbiased estimators for the location and scale parameters of the extreme value distribution using ranked set sampling. The RSS procedure was also modified to get better estimators of the parameters. In this paper we derive the estimators for the parameters of rectangular and logistic distributions by using modified ranked set sampling (MRSS) procedure.

The procedure of selection of RSS involves drawing of n random samples with n units in each sample. The n units in each sample are ranked by using judgment or by other means not requiring actual measurements. The unit with the lowest rank is quantified from the first sample, the unit with the second lowest rank is quantified from the second sample, and this procedure is continued until the unit with the highest rank is quantified from the nth sample. The n^2 ordered observations in the n samples can be displayed as:

$$\begin{matrix} x_{(11)} & x_{(12)}\cdots x_{(1n)} \\ x_{(21)} & x_{(22)}\cdots x_{(2n)} \\ \vdots & \vdots \quad\quad \vdots \\ x_{(n1)} & x_{(n2)}\cdots x_{(nn)} \end{matrix}$$

In the usual RSS procedure only n observations $x_{(11)}, x_{(22)}, \ldots, x_{(nn)}$ are accurately measured. Note that these n observations are independently distributed. Furthermore, $x_{(ii)}$ is the ith ordered observation in the ith sample. Bhoj and Ahsanullah (1996a) used these observations to derive the minimum variance linear unbiased estimators for the parameters of the rectangular distribution.

2. ESTIMATION OF PARAMETERS BASED ON RANKED SET SAMPLE

In this section we use the ranked set sample of n observations $x_{(11)}, x_{(22)}, \ldots, x_{(nn)}$ to estimate the location and scale parameters of the distribution. Let μ and σ denote the location and scale parameters of the distribution. We define

$$y_{(ii)}=(x_{(ii)}-\mu)/\sigma \ , \ E(y_{(ii)})=\alpha_i \text{ and } Var(y_{(ii)})=v_{ii}.$$

In terms of original $x_{(ii)}$'s we have

$$E(x_{(ii)})=\mu+\sigma\alpha_i \text{ and } Var(x_{(ii)})=v_{ii}\sigma^2.$$

Let $\mathbf{X}'=(x_{(11)}, x_{(22)}, \ldots, x_{(nn)})$,

$$\mathbf{1}'=(1,1,\ldots,1),$$

$$\alpha'=(\alpha_1,\alpha_2,\ldots,\alpha_n),$$

and Variance-Covariance matrix of $\mathbf{X}=V\sigma^2$, where V is an nxn diagonal matrix with v_{ii} as $(i,i)^{th}$ element. Then we can write

$$E(\mathbf{X}) = \mu\mathbf{1}+\sigma\alpha$$
$$= A\theta \quad ,$$

where $A'=\begin{pmatrix} 1 & 1 & \ldots & 1 \\ \alpha_1 & \alpha_2 & \ldots & \alpha_n \end{pmatrix}$ and $\theta'=(\mu,\sigma)$.

The minimum variance linear unbiased estimators (MVLUE) of θ is obtained by the least squares theorem of Gauss and Markoff. If

$\hat{\theta}$ denotes the MVLUE of θ, then $\hat{\theta}=(A'V^{-1}A)^{-1}A'V^{-1}X$.

After some simplifications, we can write

$$\hat{\mu}=\sum_{i=1}^{n} w_{1i}X_{(ii)} \tag{2.1}$$

$$\hat{\sigma}=\sum_{i=1}^{n} w_{2i}X_{(ii)} \quad , \tag{2.2}$$

where

$$w_{1i}=(T_1-\alpha_i T_3)/(Dv_{ii}) \quad ,$$

$$w_{2i}=(\alpha_i T_2-T_3)/(Dv_{ii}) \quad ,$$

$$T_1=\sum_{i=1}^{n}(\alpha_i^2/v_{ii}) \quad , \quad T_2=\sum_{i=1}^{n}(1/v_{ii}) \quad , \tag{2.3}$$

$$T_3=\sum_{i=1}^{n}(\alpha_i/v_{ii}) \text{ and } D=T_1T_2-T_3^2 . \tag{2.4}$$

The variances and covariance of these estimators are given by

$$Var(\hat{\mu})=\frac{\sigma^2 T_1}{D} \quad , \quad Var(\hat{\sigma})=\frac{\sigma^2 T_2}{D} \text{ and} \tag{2.5}$$

$$Cov(\hat{\mu},\hat{\sigma})=-\frac{\sigma^2 T_3}{D} . \tag{2.6}$$

This method has been used to estimate parameters of various distributions. In the next sections, the variances of the estimators based on RSS are compared with those of ordered least squares estimators of two distributions. The RSS procedure, as will be shown in the next sections, does not work well for estimating particularly the scale parameter when n is small. Therefore, in this section, we outline the modification of the RSS procedure which will further reduce the variances of both location and scale parameters. For this purpose we have to assume that n(n=2m) is even.

In MRSS, we select only two appropriate order statistics instead of n order statistics as is done in the RSS. We select m j^{th} order statistic and m k^{th} order statistic. For convenience, we assume that the jth order statistic is selected from the first m samples and the kth order statistic is selected from the last m samples. The choices of jth and k^{th} order statistics depend upon the distribution under consideration and the parameter(s) to be estimated. These choices will be explained in the next sections with reference to two symmetric distributions.

3. RECTANGULAR DISTRIBUTION

The probability density function of the rectangular distribution centered at μ with variance σ^2 is given by

$$f(x)=\frac{1}{2\sqrt{3}\sigma} \quad , \quad \mu-\sqrt{3}\sigma\leq x\leq\mu+\sqrt{3}\sigma$$

$$=0 \quad , \text{ otherwise.}$$

The best linear unbiased estimators for μ and σ, given by Downton (1954) by using Lloyd's (1952) method, are based only on the largest and smallest ordered observations with variances and covariance given by

$$Var(\hat{\mu})=\frac{6\sigma^2}{(n+1)(n+2)}, \quad Var(\hat{\sigma})=\frac{2\sigma^2}{(n-1)(n+2)} \quad \text{and } Cov(\hat{\mu},\hat{\sigma})=0.$$

Bhoj and Ahsanullah (1996a) derived the minimum variance linear unbiased estimators for μ and σ based on RSS and they are given by

$$\tilde{\mu}=\sum_{i=1}^{n} w_{1i}x_{(ii)} \quad , \quad \tilde{\sigma}=\sum_{i=1}^{n} w_{2i}x_{(ii)} \quad , \tag{3.1}$$

where

$$w_{1i}=\frac{1}{v_{ii}T_1}=\frac{n+1}{2i(n-i+1)s_n} \quad ,$$

$$w_{2i}=\frac{\alpha_i}{v_{ii}T_2}=\frac{(n+1)(2i-n-1)}{2\sqrt{3}i(n-i+1)\{(n+1)s_n-2n\}} \quad ,$$

$$\alpha_i=\sqrt{3}\left\{\frac{2i}{n+1}-1\right\} \quad , \quad v_{ii}=\frac{12i(n-i+1)}{(n+1)^2(n+2)} \quad , \quad s_n=\sum_{i=1}^{n}\frac{1}{i} \tag{3.2}$$

and T_1 and T_2 are given in (2.3).

w_{1i} and w_{2i} satisfy the following equations:

$$\Sigma w_{1i}=1 \quad , \quad \Sigma w_{1i}\alpha_i=0 \quad , \quad \Sigma w_{2i}=0 \quad \text{and} \quad \Sigma w_{2i}\alpha_i=1.$$

The variances and covariance of $\tilde{\mu}$ and $\tilde{\sigma}$ are given by

$$Var(\tilde{\mu})=\frac{\sigma^2}{T_2}=\frac{6\sigma^2}{(n+1)(n+2)s_n} \tag{3.3}$$

$$Var(\tilde{\sigma})=\frac{\sigma^2}{T_1}=\frac{2\sigma^2}{(n+2)\{(n+1)s_n-2n\}}\ ,\ Cov(\tilde{\mu},\tilde{\sigma})=0. \qquad (3.4)$$

It is clear that $Var(\tilde{\mu})<Var(\hat{\mu})$ for all n since $s_n>1$. However, $Var(\tilde{\sigma})<Var(\hat{\sigma})$ when n>5. Thus the RSS method does not work well for estimating σ since it was devised primarily for estimating population mean. We note that

$$\alpha_i=-\alpha_{n-i+1}\ ,\ i\leq[(n+1)/2]$$

with $\alpha_{[n/2]+1}=0$ for odd n, where [·] denotes the largest integer contained in [·]. $\alpha_i<0$ for $i\leq[(n+1)/2]$. We also observe that $v_{ii}=v_{n-i+1\ n-i+1}$. The values $|\alpha_i|$ are the largest when i=1 and n, and the values of v_{ii} are the smallest for i=1 and n. Therefore for i=1 or n, the values of T_1 and T_2 are the largest and the variances of the estimators will be the minimum. Then our modified RSS procedure is to use half smallest order statistics and half largest order statistics. In writing the formulae for the estimators we use the smallest order statistics from the first m samples and the largest order statistics from the last m samples. Let $\mu*$ and $\sigma*$ denote the estimators of μ and σ with modified RSS procedure. Then

$$\mu*=\left[\sum_{i=1}^{m}x_{(i1)}+\sum_{i=m+1}^{n}x_{(in)}\right]/n,$$

$$\sigma*=\left[(\alpha_1/v_{11})\sum_{i=1}^{m}x_{(i1)}+(\alpha_n/v_{nn})\sum_{i=m+1}^{n}x_{(in)}\right]/[n(\alpha_n^2/v_{nn})].$$

The variances of μ^* and σ^* are given by

$$Var(\mu^*)=\frac{12\sigma^2}{(n+1)^2(n+2)} \text{ and } Var(\sigma^*)=\frac{4\sigma^2}{(n-1)^2(n+2)}.$$

In the next section we compare various estimators of μ and σ.

4. COMPARISONS OF ESTIMATORS OF RECTANGULAR DISTRIBUTION

The usual estimator, proposed by McIntyre (1952), for the population mean based on RSS is $\bar{\mu}=\sum_{i=1}^{n} x_{(ii)}/n$ with the variance

$Var(\bar{\mu})=2\sigma^2/(n(n+1))$. It is known that relative efficiency of $\bar{\mu}$ with respect to the sample mean of SRS is maximum when the underlying distribution is rectangular. In this section we compare our estimators μ^* and σ^* with the known estimators $\hat{\mu}$, $\tilde{\mu}$, $\bar{\mu}$, $\hat{\sigma}$ and $\tilde{\sigma}$ by computing the following relative precisions:

$$RP_1=\frac{Var(\hat{\mu})}{Var(\mu^*)}=\frac{n+1}{2}, \quad RP_2=\frac{Var(\tilde{\mu})}{Var(\mu^*)}=\frac{(n+1)}{2s_n},$$

$$RP_3=\frac{Var(\bar{\mu})}{Var(\mu^*)}=\frac{(n+1)(n+2)}{6n}, \quad RP_4=\frac{Var(\hat{\sigma})}{Var(\sigma^*)}=\frac{n-1}{2},$$

and $$RP_5=\frac{Var(\tilde{\sigma})}{Var(\sigma^*)}=\frac{(n-1)^2}{2\{(n+1)s_n-2n\}}.$$

We observe that RP_1 is exactly the relative precision of $\bar{\mu}$ with respect to the sample mean of SRS. The numerical computations of the relative precisions are presented in Table 1 for n=2(2)14.

We note that $\mu*$ has substantial gain over $\hat{\mu}$ for all n, and $\mu*$ is better than $\hat{\mu}$ and $\bar{\mu}$ for n>2. $\sigma*$ is superior to $\hat{\sigma}$ for n≥4, and it is better than $\tilde{\sigma}$ for n>2. Again the gains in precisions are substantial.

Table 1. Relative precisions of $\mu*$ and $\sigma*$ for rectangular distribution.

n	RP_1	RP_2	RP_3	RP_4	RP_5
2	1.5000	1.0000	1.0000	0.5000	1.0000
4	2.5000	1.2000	1.2500	1.5000	1.8621
6	3.5000	1.4286	1.5556	2.5000	2.4272
8	4.5000	1.6557	1.8750	3.5000	2.8957
10	5.5000	1.8778	2.2000	4.5000	3.3146
12	6.5000	2.0946	2.5278	5.5000	3.7022
14	7.5000	2.3066	2.8571	6.5000	4.0677

5. ESTIMATION OF EXTREMITIES OF RECTANGULAR DISTRIBUTION

In the case of the rectangular distribution, it may be more natural to estimate the extremities $\lambda_1=\mu-\sqrt{3}\sigma$ and $\lambda_2=\mu+\sqrt{3}\sigma$ rather than the mean and the standard deviation. The estimators of λ_1 and λ_2 based on ordered observations can be obtained simply by using the least squares estimators of μ and σ; see Lloyd (1952). In this section we compare the variances of the three estimators of λ_1 and λ_2. The three estimators are

$$\hat{\lambda}_1=\hat{\mu}-\sqrt{3}\hat{\sigma} \quad , \quad \hat{\lambda}_2=\hat{\mu}+\sqrt{3}\hat{\sigma}, \tag{5.1}$$

$$\tilde{\lambda}_1=\tilde{\mu}-\sqrt{3}\tilde{\sigma} \quad , \quad \tilde{\lambda}_2=\tilde{\mu}+\sqrt{3}\tilde{\sigma}, \tag{5.2}$$

$$\lambda_1^*=\mu^*-\sqrt{3}\sigma^* \quad , \quad \lambda_2^*=\mu^*+\sqrt{3}\sigma^*, \tag{5.3}$$

where (5.1), (5.2) and (5.3) are respectively obtained from the ordered least squares estimators, RSS estimators and MRSS estimators. Since all three estimators of μ and σ are uncorrelated, the variances of both extremities are equal. The variances of the estimators of the extremities are given by

$$Var(\hat{\lambda}_1) = \frac{12n\sigma^2}{(n^2-1)(n+2)} ,$$

$$Var(\tilde{\lambda}_1) = \frac{12n(s_n-1)\sigma^2}{(n+2)(n-1)S_n\{(n+1)s_n-2n\}} ,$$

and $$Var(\lambda_1^*) = \frac{24(n^2+1)\sigma^2}{(n-1)^2(n+1)^2(n+2)} .$$

The three estimators of the extremities are compared by computing the following three relative precisions:

$$RPE_1 = \frac{Var(\hat{\lambda}_1)}{Var(\tilde{\lambda}_1)} , \quad RPE_2 = \frac{Var(\hat{\lambda}_1)}{Var(\lambda_1^*)} \text{ and } RPE_3 = \frac{Var(\tilde{\lambda}_1)}{Var(\lambda_1^*)} .$$

Table 2 presents the numerical computations of the variances and the relative precisions of the three estimators. We observe that $\hat{\lambda}_1$ is better than the other two estimators only for n=2. When n>2, the estimator λ_1^* is superior to $\hat{\lambda}_1$ and $\tilde{\lambda}_1$ with substantial gains in the relative precisions.

Table 2. Variances and relative precisions of extremities.

n	$\frac{Var(\hat{\lambda}_1)}{\sigma^2}$	$\frac{Var(\tilde{\lambda}_1)}{\sigma^2}$	$\frac{Var(\lambda_1^*)}{\sigma^2}$	RPE_1	RPE_2	RPE_3
2	2.0000	3.3333	3.3333	0.6000	0.6000	1.0000
4	0.5333	0.5098	0.3022	1.0462	1.7647	1.6868
6	0.2571	0.1894	0.0906	1.3579	2.8378	2.0898
8	0.1524	0.0954	0.0393	1.5965	3.8769	2.4283
10	0.1010	0.0564	0.0206	1.7897	4.9010	2.7385
12	0.0719	0.0368	0.0122	1.9520	5.9172	3.0314
14	0.0538	0.0257	0.0078	2.0919	6.9289	3.3123

6. LOGISTIC DISTRIBUTION

The random variable X has a logistic distribution if it has a probability density function (pdf) of the form:

$$f(x) = \frac{\pi}{\sigma\sqrt{3}} \frac{e^{-\pi(x-\mu)/(\sqrt{3}\sigma)}}{\{1+e^{-\pi(x-\mu)/(\sqrt{3}\sigma)}\}^2}, \quad -\infty<x,\mu<\infty, \quad \sigma>0, \tag{6.1}$$

$$= 0 \quad \text{otherwise,}$$

where μ and σ are the mean and standard deviation of the distribution. Lam, Sinha and Wu (1995) have taken a slightly different form of the pdf of logistic distribution. They obtained the best linear unbiased estimators of the location, scale parameters and the quantiles of the logistic distribution. They compared the variances of these estimators with those based on SRS and Cramer-Rao lower bound. Similar work has been carried out by Bhoj and Ahsanullah (1996b) for estimation of the mean and standard deviation of the logistic distribution given by (6.1). They compared the variances of the estimators of μ and σ with the estimators obtained from the ordered least squares estimators. They have also computed the coefficients w_{1i} and w_{2i} given by (2.1) and (2.2) for the logistic distribution. In this paper, we obtain the estimators of μ and σ based on MRSS and compare those with the known estimators.

For the logistic distribution α_i and v_{ii} as defined in Section 2 are given by

$$\alpha_i = \frac{\sqrt{3}}{\pi}\{\psi(i) - \psi(n-i+1)\}$$

$$v_{ii} = \frac{3}{\pi^2}\{\psi'(i) + \psi'(n-i+1)\} \quad , \quad i=1,2,\ldots,n,$$

where $\psi(z) = \frac{d}{dz}\ln\Gamma(z)$ and $\psi'(z) = \frac{d^2}{dz^2}\ln\Gamma(z)$ are the digamma and trigamma functions respectively. α_i and v_{ii} can be computed by using the algorithms given by Bernado (1976) and Schneider (1978). The estimators of μ and σ based on RSS are denoted by $\tilde{\mu}$ and $\tilde{\sigma}$ as in the previous sections. The variances of $\tilde{\mu}$ and $\tilde{\sigma}$ are computed from (2.5) and it can be shown that covariance between $\tilde{\mu}$ and $\tilde{\sigma}$ is zero. Best linear unbiased estimators for μ and σ of the logistic distribution using order statistics are given by Gupta, Qureishi and Shah (1967) by using the method of Lloyd (1952). As before, we denote these estimators by $\hat{\mu}$ and $\hat{\sigma}$. For comparison purposes the variances of $\hat{\mu}$ and $\hat{\sigma}$ are taken from Balakrishnan and Leung (1988). It is known that the covariance between $\hat{\mu}$ and $\hat{\sigma}$ is zero. It is known that $\hat{\mu}$ is uniformly better than $\tilde{\mu}$ and $\tilde{\sigma}$ is better than $\hat{\sigma}$ for n>4. Furthermore, the gain in precision in estimating the population mean is quite substantial.

7. ESTIMATION OF μ AND σ USING MRSS

In this section, we use MRSS to estimate the mean and standard deviation of the logistic distribution. The properties of α_i and v_{ii} of the logistic distribution are the same as those of the rectangular distribution described in section 3. If we wish to estimate μ then the best choice is to select the median from each sample and take the average of these medians when n is odd. If n is even and we choose the median of order statistics from each sample, then we will be quantifying 2n observations. The comparison of this estimator will not be fair with the other estimators which are based on only n observations. Therefore, for n=2m, we choose the mth order statistic from the first half samples and $(m+1)^{th}$ order statistic from the last half samples. Thus our estimator for the mean is

$$\mu^* = \left[\sum_{i=1}^{m} x_{(im)} + \sum_{i=m+1}^{n} x_{(i,m+1)}\right] / n \ ,$$

with variance

$$Var(\mu^*) = (n/v_{kk})^{-1}\sigma^2 \ , \quad \text{where } k = [\frac{n+1}{2}] .$$

Let σ^* denote the estimator for σ with MRSS procedure. The values of α_i and v_{ii} suggest to use the extreme order statistics from each sample (set). However, as n increases, the choice of the appropriate order statistics changes. Table 3 gives the appropriate order statistics for even sample sizes up to 30.

Table 3. Appropriate order statistics for σ^*.

Range of n	2 - 12	14 - 24	26 - 30
Order statistics	1 , n	2 , (n-1)	3 , (n-3)

The estimator σ^* is given by

$$\sigma^* = \frac{(\alpha_j/v_{jj}) \sum_{i=1}^{m} x_{(ij)} + (\alpha_k/v_{kk}) \sum_{i=m+1}^{n} x_{(ik)}}{n(\alpha_k^2/v_{kk})},$$

where j=1 and k=n for $n \leq 12$

and j=2 and k=n-1 for $14 \leq n \leq 24$.

The variance of σ^* is

$$Var(\sigma^*) = (n\alpha_k^2/v_{kk})^{-1}\sigma^2.$$

Now we compare the estimators μ^* and σ^* with the other estimators of μ and σ discussed in section 6. μ^* is also compared with the usual estimator based on RSS. This estimator is $\overline{\mu}=\Sigma x_{(ii)}/n$ with the variance $Var(\overline{\mu}) = (\sigma^2/n^2)\Sigma v_{ii}$. For comparison purposes we compute the same five relative precisions as defined in section 4. These relative precisions along with the variances of μ^* and σ^* are given in Table 4 for n=2(2)14.

We observe that the relative precision of μ^* over $\hat{\mu}$ is dramatically higher for all values of n. The relative precisions of μ^* over $\hat{\mu}$ and $\overline{\mu}$ are higher than one when n>2. We note that σ^* is superior to $\tilde{\sigma}$ and $\hat{\sigma}$ when n>2. The relative precision of σ^* over $\hat{\sigma}$ increases with n. However, the relative precision of σ^* over $\tilde{\sigma}$ increases with n up to n=8 and then decreases until n=12 at which point we cease to use the order statistics 1 and n.

Table 4. Variances and relative precisions of $\mu*$ and $\sigma*$ for logistic distribution.

n	$\frac{Var(\mu*)}{\sigma^2}$	RP_1	RP_2	RP_3	$\frac{Var(\sigma*)}{\sigma^2}$	RP_4	RP_5
2	0.3480	4.7266	1.0000	1.0000	1.1449	0.5633	1.0000
4	0.0790	10.0340	1.2994	1.4274	0.1435	1.5711	1.7575
6	0.0344	15.1215	1.3920	1.6556	0.0584	2.3461	1.9583
8	0.0192	20.1368	1.4320	1.8103	0.0331	2.9767	1.9893
10	0.0122	25.1198	1.4530	1.9276	0.0219	3.5120	1.9586
12	0.0085	30.0873	1.4654	2.0220	0.0158	3.9791	1.9045
14	0.0062	35.0462	1.4733	2.1012	0.0117	4.5587	1.9110

REFERENCES

Balakrishnan, N. and Leung, M.Y. (1988). Means, variances and covariances of order statistics, BLUE's for the type I generalized logistic distribution, and some applications. Communications in Statistics, Simulation and Computation, 17, 51-84.

Bernardo, J.M. (1976). Psi (digamma) function. Algorithm AS 103, Applied Statistics, 25, 315-317.

Bhoj, D.S. (1996a). Estimation of mean and standard deviation of Laplace and normal distributions using ranked set sampling. Submitted for publication.

Bhoj, D.S. (1996b). Estimation of parameters of the extreme value distribution using ranked set sampling. In press.

Bhoj, D.S. and Ahsanullah, M. (1996a). Estimation of parameters of the generalized geometric distribution using ranked set sampling. Biometrics, 52, 685-694.

Bhoj, D.S. and Ahsanullah, M. (1996b). Estimation of the parameters of the logistic distribution using ranked set sampling. Unpublished manuscript.

Bohn, L.L. and Wolfe, D.A. (1992). Nonparametric two-sample procedures for ranked-set samples data. Journal of the American Statistical Association, 87, 552-561.

David, H.A. and Levine, D.N. (1972). Ranked set sampling in the presence of judgment error. Biometrics, 28, 553-555.

Dell, T.R. and Clutter, J.L. (1972). Ranked-set sampling theory with order statistics background. Biometrics, 28, 545-555.

Downton, F. (1954). Least-squares estimates using ordered observations. Annals of Mathematical Statistics, 25, 303-316.

Gupta, S.S., Qureishi, A.S. and Shah, B.K. (1967). Best linear unbiased estimators of the parameters of the logistic distribution using order statistics, Technometrics, 9, 43-56.

Halls, L.K. and Dell, T.R. (1996). Trial of ranked set sampling for forage yields. Forest Science, 12, 22-26.

Lam, K., Sinha, B.K. and Wu Zhong (1995). Estimation of location and scale parameters of a logistic distribution using a ranked set sample. In collected essays in honor of Professor Hubert A. David, 189-197.

Lloyd, E.H. (1952). Least-squares estimation of location and scale parameters using order statistics. Biometrika, 39, 88-95.

McIntyre, G.A. (1952). A method of unbiased selective sampling, using ranked sets. Australian Journal of Agricultural Research, 3, 385-390.

Muttlak, H.A. (1995). Parameter estimation in linear regression using ranked set sampling. Biometrical Journal, 37, 799-811.

Muttlak, H.A. and McDonald, L.L. (1990). Ranked set sampling with size-biased probability of selection. Biometrics, 46, 435-445.

Patil, G.P., Sinha, A.K. and Taillie, C. (1995). Finite population corrections for ranked set sampling. Annals of the Institute of Statistical Mathematics, 47, 621-636.

Patil, G.P., Sinha, A.K. and Taillie, C. (1993a). Ranked set sampling from a finite population in the presence of a trend on a site. Journal of Applied Statistical Science, 1, 51-65.

Patil, G.P., Sinha, A.K. and Taillie, C. (1993b). Relative precision of ranked set sampling: A Comparison with the regression estimator. Environmetrics; 4, 399-412.

Schneider, B.E. (1978). Trigamma function. Algorithm AS 121, Applied Statistics, 27, 97-99.

Stokes, S.L. (1995). Parametric ranked set sampling. Annals of the Institute of Statistical Mathematics, 47, 465-482.

Stokes, S.L. (1980). Estimation of variance using judgement ordered ranked-set samples. Biometrics, 36, 35-42.

Stokes, S.L. and Sager, T.W. (1988). Characterization of a ranked-set sample with application to estimating distribution functions. Journal of the American Statistical Association, 83, 374-381.

Stokes, S.L. (1977). Ranked set sampling with concomitant variables. Communications in Statistics - Theory and Methods, A6(12), 1207-1211.

Takahasi, K. and Wakimoto, K. (1968). On unbiased estimates of the population mean based on the sample stratified by means of ordering. Annals of the Institute of Statistical Mathematics, 20, 1-31.

Yanagawa, T. and Shirahata, S. (1976). Ranked set sampling theory with selective probability matrix. Australian Journal of Statistics, 18, 45-52.

Applied Statistical Science, II
ISBN 1-56072-469-2

CERTAIN FAMILIES OF DISTRIBUTIONS, INDEXED BY α.

A. Reza Soltani*
Department of Statistics
Shiraz University

September 30, 1996

Abstract

In this work, by considering certain random vectors on the curve $x^\alpha + y^\alpha = 1$, we derive new families of distribution functions indexed by $1 \le \alpha \le 2$. The triangle of Uniform distribution, the $\beta(1/2, 1/2)$, , and the Arcsin distribution on (0,1) is furnished by the presented families. Some issues concerning α-symmetric distributions are discussed.

1. Introduction

Let $D_\alpha^+ = \{(x_1, x_2) : x_1^\alpha + x_2^\alpha = 1,\ 0 \le x_1 \le 1,\ 0 \le x_2 \le 1\}$, $1 \le \alpha \le 2$. In this paper we consider certain bivariate random vectors (X, Y) supported by D_α^+, and then study the marginal distribution of X. This leads to the

*This research was supported by the Shiraz University Research Council

derivation of certain distributions on $(0,1)$ that are indexed by $\alpha, 1 \leq \alpha \leq 2$. More precisely we derive the following density functions.

$$
\begin{aligned}
f_\alpha(x) &= \frac{\alpha}{\pi}(1-x^\alpha)^{-1/2}x^{\frac{\alpha}{2}-1}, \ 0 \leq x \leq 1, \\
g_\alpha(x) &= \ell_\alpha\{1+x^{2(\alpha-1)}(1-x^\alpha)^{\frac{2}{\alpha}-2}\}^{1/2}, \ 0 \leq x \leq 1 \\
h_\alpha(x) &= \ell_\alpha\{x^{\frac{2}{\alpha}-2}+(1-x)^{\frac{2}{\alpha}-2}\}^{1/2}, \ 0 < x < 1,
\end{aligned}
$$

where l_α is the normalizing constant.

Interestingly,

f_1 is the density of beta (1/2, 1/2),

f_2 is the density of arcsin.

Thus beta (1/2, 1/2) and arcsin laws are border members of the family $\{f_\alpha, \ 1 \leq \alpha \leq 2\}$. Also, note that

g_1 is the density of uniform on (0,1),

g_2 is the density of arcsin on (0,1).

Therefore uniform and arcsin laws are furnished by $\{g_\alpha, \ 1 \leq \alpha \leq 2\}$. Also,

h_1 is the density of uniform on (0,1),

h_2 is the density of bet (1/2, 1/2).

So the family $\{h_\alpha, 1 \leq \alpha \leq 2\}$ is started by uniform and is terminated by beta(1/2,1/2).

2 The Densities

In this section we derive the densities f_α, g_α and h_α, and provide a relationship between them.

2.1 Lemma. Let (X, Y) be uniformly distributed on the first quarter of the unit circle, then X possesses arcsin law on (0,1).

Proof .

$$P(X \leq x) = P(\theta(X,Y) > Arccos(x)) = 1 - \frac{2}{\pi} Arccos(x)$$

2.2 Lemma. Let (X,Y) be uniformly distributed on D_2^+ then $(X^{2/\alpha}, Y^{2/\alpha})$ is distributed on D_α^+ and $X^{2/\alpha}$ has the density f_α.

Proof. This is immediate by using the transformation formula.

2.3 Lemma. Suppose (X,Y) are distributed on D_α^+ according to the normalized Lebesgue measure, then X has the density g_α.

Proof. Note that

$$P(X \leq x) = \ell_\alpha \int_0^x \sqrt{1 + y'^2}\, dt,$$

where $y^\alpha + x^\alpha = 1$.

2.4 Lemma. Suppose X_α possesses the density g_α, then X_α^α has density h_α.

Proof. This is immediate by using the transformation formula.

2.5 Notations. The symbol $X \frown S(\alpha)$ stands for a random variable X having density g_α, while $X \frown R(\alpha)$ indicates that X has density h_α.

2.6 Remark. If $X \frown S(\alpha)$, then $(1 - X^\alpha)^{1/\alpha} \frown S(\alpha)$. If $Y \frown R(\alpha)$, then

1- $Y \frown R(\alpha)$. Therefore

$$EY = EX^{\alpha} = \frac{1}{2}$$

2.7 Remark. We do not have an explicit formula for the ℓ_{α}, but in terms of infinite series the following formula can be obtained,

$$\ell_{\alpha} = \left(1 + \sum_{n=1}^{\infty} \frac{(-1)^{n+1} \prod_{k=0}^{n-1}(\alpha - 2k)}{2^{n}(n+1)!}\right)^{-1}$$

2.8 Remark. If $(X,Y)_{\alpha}$ is distributed on $D_{\alpha} = \{(x_1,x_2) : |x_1|^{\alpha} + |x_2|^{\alpha} = 1\}$, $1 \leq \alpha \leq 2$, according to the normalized Lebesgue measure, then its characteristic function is given by

$$\begin{aligned} \psi_{\alpha}(t,s) &= Ee^{i(tX+sY)} \\ &= \int_0^1 cos(tx)cos(s(1-x^{\alpha})^{1/\alpha})\, g_{\alpha}(x)dx. \end{aligned}$$

Note that

$$\psi_1(t,s) = \frac{sin(t)+sin(s)}{t+s} + \frac{sin(t)-sin(s)}{t-s},$$

$$\psi_2(t,s) = \int_0^{\pi/2} cos(rcos(x))dx, \quad r = t^2 + s^2.$$

Therefore $(X,Y)_1$ is not 1-symmetric, but as it was noted by Shoenberg $(X,Y)_2$ is 2-symmetric. The figure of ψ_{α} indicates, $(X,Y)_{\alpha}$, $1 < \alpha < 2$, is not α-symmetric, but the range of the values of ψ_{α} on $t^{\alpha} + s^{\alpha} = 1$ is considerably small.

Acknowledgment. The Author is greateful to N. Abbasi for providing the formula for ℓ_{α} in Remark 2.7.

References

Cambanis, S. Keener, R. and Simon, G. (1984). On α- symmetric multivariate distributions. J. Multivariate Anal. 13 213-233.

Schoenberg, I. J. (1938). Metric spaces and completely monotone functions. Ann. Math. 39 811-841.

INDEX